Relic, Icon or Hoax?
Carbon Dating the Turin Shroud

Relic, Icon or Hoax?
Carbon Dating the Turin Shroud

H E Gove

Professor Emeritus of Physics
University of Rochester

With a Foreword by D Allan Bromley

Institute of Physics Publishing
Bristol and Philadelphia

British Library Cataloguing-in-Publication Data
A catalogue record for this book is available from the British Library.

ISBN 0 7503 0398 0

Library of Congress Cataloging-in-Publication Data
Gove, H. E. (Harry Edmund), 1922–
    Relic, icon or hoax? : carbon dating the Turin shroud / H. E. Gove ; with a foreword by D. Allan Bromley.
      p.   cm.
    Includes index.
    ISBN 0-7503-0398-0 (alk. paper)
    1. Holy shroud.   I. Title.
    BT587.S4G68 1996
    232.96'6 - - dc20   96-30321
                                                    CIP

Published by Institute of Physics Publishing, wholly owned by The Institute of Physics, London

Institute of Physics Publishing, Techno House, Redcliffe Way, Bristol BS1 6NX, UK
US Editorial Office: Institute of Physics Publishing, Suite 1035, The Public Ledger Building, 150 South Independence Mall West, Philadelphia, PA 19106, USA

Typeset in TEX using the IOP Bookmaker Macros
Printed and bound in Great Britain at the University Press, Cambridge

To Shirley

Contents

Foreword

The relationship between religion and science has always been somewhat uneasy, at times hostile, and one frequently marked by misunderstandings—both minor and major.

The potential for such misunderstandings—and hostility—increases dramatically when fundamental science and technology are applied to one of the most famous and revered relics in all of Christendom—the Shroud of Turin.

First entering the historical record in 1353 when a French knight, Geoffroi I de Charney, placed it in a church in Lirey, France, the shroud has been widely believed to be the burial shroud of the crucified Jesus Christ. Passing from de Charney's daughter to the House of Savoy, it was exhibited in Chambery, France, and finally, in 1578, was moved to Turin where it remains today.

Despite numerous scientific—and quasi-scientific—studies, the source of the striking negative image on the shroud remains a mystery, and from its earliest appearance there have always been questions as to the shroud's authenticity. With the discovery of carbon dating, by Willard Libby in the 1940s, it was clear that establishing the date on which the flax from which the Shroud was woven had been harvested was possible in principle. Such dating, were it to establish an age around 2000 years, would be consistent with, but could not prove, the shroud's authenticity; a substantially shorter age would, however, prove that it was not authentic. Unfortunately, Libby's dating technique would have required the destruction of an unacceptably large area of the shroud and this, together with an understandable reluctance on the part of its religious custodians to have its age established at all, precluded any attempt at dating.

All this changed in May of 1977 when Harry Gove, Ted Litherland and Kenneth Purser, at Gove's tandem accelerator laboratory at the University of Rochester, developed so-called accelerator mass spectrometry and reduced the sample size required by factors between 100 and 1000. Among the consequences of the widespread publicity accorded this development was a June 1977 letter to Gove from an English cleric, suggesting that at last the dating of the Shroud of Turin was entirely feasible. Gove was intrigued, not because of any strong religious biases but because he recognized that dating the shroud would be an excellent demonstration of the power of his new technology. While technological feasibility seemed obvious, little did he suspect the non-technical difficulties that lay ahead.

As a long-time colleague of Gove's, in experimental nuclear physics, I may be permitted a slight digression to emphasize that accelerator mass spectrometry is yet another in a long list of developments originating in nuclear science and technology that have revolutionized fields of science from cosmology to archeology and of technology from microelectronics to clinical medicine. Gove has played a key role in a number of them.

This book is the remarkable record of the above-mentioned difficulties, over more than a decade, and involving a cast of characters including Pope John Paul II, Senator Moynihan of New York, the President of the Papal Academy of Sciences, the Archbishop of Turin, the personnel of some seven laboratories in Europe and the US equipped for accelerator mass spectrometry, an international cast of journalists and what must have been an astronomical telephone bill!

The record includes deception, outright lies, low cunning, misrepresentation, and a pathological hunger for publicity as well as solid science and technology, faith that passeth all understanding, and, on Gove's part, tenacity and determination rarely encountered.

As in any good suspense story a somewhat mysterious group, the Shroud of Turin Research Project (or STURP) lurks in the background. Having made extensive earlier measurements on the shroud using a wide variety of physical measurement techniques—but learning relatively little from them—the STURP group attempted to block the carbon dating of the shroud and, when that failed, attempted to move in and take credit for as much of the subsequent activity as possible. The STURP members, as well as the science advisor to the archbishop of Turin, did little for the reputations of science or scientists!

Suffice it to say that in the end the age of the shroud was

established—but not in the way or by the people that would have been expected. This is a fascinating and unusual book; it is a very personal memoir; and it provides a rare window into the sometimes surprising workings of both science and religion.

And through it all, Gove's original intent—that of demonstrating that accelerator mass spectrometry has the potential to revolutionize whole areas of science—was, indeed, fulfilled.

D Allan Bromley
New Haven, Connecticut

Acknowledgments

The Turin Shroud would not have been carbon dated were it not for the technology of accelerator mass spectrometry (AMS). That technology was invented and initially developed principally as a result of a collaboration between scientists at the University of Toronto led by Professor A E Litherland, at the General Ionex Corporation led by Dr K H Purser and at the University of Rochester led by the author. I wish to express my gratitude and indebtedness to my two colleagues Ted Litherland and Ken Purser who made and continue to make major contributions to the art and science of AMS. Our collaboration has continued through the years and is a source of great pleasure to me. The three of us benefited greatly from our association with Dr Meyer Rubin, head of the US Geological Survey's Radiocarbon Laboratory in Reston, Virginia. He taught us much about carbon dating and participated in the early work in AMS at the University of Rochester.

The actual AMS measurements were carried out at facilities at Arizona, USA, Oxford, England and Zurich, Switzerland. It is to the credit of the scientists at those three facilities that everything worked so well and that such excellent agreement was obtained from totally independent measurements. I am particularly grateful to Professor D J Donahue, co-director of the NSF Accelerator Facility for Radioisotope Analysis at the University of Arizona for inviting me to be present when the first AMS measurement of the Turin Shroud's age was made. Throughout the complex and often frustrating interactions that led to the final measurements, he always acted wisely and correctly.

This curious association between science and religion that led to the dating of the Turin Shroud succeeded because of the participation of the Pontifical Academy of Sciences led by its president Professor Carlos Chagas. In this effort Professor Chagas was ably assisted by Dr Vittorio Canuto of the NASA Institute for Space Studies in New York City. The efforts of Professor Chagas and Dr Canuto were crucial to the eventual success of the project. Professor Chagas is the exemplar of a scholar and a gentleman. He retired as president of the academy in 1988 and it is doubtful whether, without him, the academy will be as liberal and progressive.

It gives me great pleasure to thank my daughter Pauline L Gove for reading an early version and parts of the latest version of this book and for making many valuable comments and suggestions. Dorothy Crispino, editor of *Shroud Spectrum International* published in Nashville, Indiana, kindly agreed to read a recent draft of the book. Her many editorial suggestions were extremely helpful and I thank her for them. Another person who cast a critical eye on the book was my lawyer and long-time friend Leon Katzen. His advice on tempering some of my more outspoken passages was invaluable. Another long-time friend and scientific colleague D Allan Bromley kindly accepted my invitation to write the foreword. His contributions to science and engineering in this country and, indeed, the world are legendary.

Finally I would like to thank Shirley Brignall, to whom the book is dedicated, for help and support in too many ways to enumerate here and which led, finally, to getting the book published.

Harry Gove
May 1996

Brief Biographical Sketch of the Author

Harry E Gove

Harry Gove was born in Niagara Falls, Ontario, Canada. He received his BSc degree in Engineering Physics at Queen's University in Kingston, Ontario, Canada in 1944 and his PhD degree from the Massachusetts Institute of Technology in Nuclear Physics in 1950. He served as Branch Head of Nuclear Physics at Atomic Energy of Canada Ltd, Chalk River, from 1956 to 1963 and in 1963 he was appointed as a Professor of Physics at the University of Rochester. He became a US citizen in 1969.

He directed the Nuclear Structure Research Laboratory at the University of Rochester from 1963 to 1988 and was chair of the Department of Physics and Astronomy from 1977 to 1980. He became Professor Emeritus of Physics in 1992. He was on leave at the Niels Bohr Institute in Copenhagen, Denmark, in 1961–62, at the Centre for Nuclear Research in Strasbourg, France, in 1971–72 and at Worcester College, Oxford, England, in 1983–84 where he was the R T French Visiting Professor. He has served on numerous visiting and advisory committees in the USA and Canada. He was a member of the Board of Trustees of Associated Universities, Inc. from 1978 to 1983. He was the Nuclear Physics Division Associate Editor of *Physical Review Letters* from 1975 to 1979 and Associate Editor of *Annual Reviews of Nuclear and Particle Physics* from 1978 to 1994. He was the recipient of the JARI (*Journal of Applied Radiation and Instrumentation*) Award from Pergamon Press for outstanding contributions to the development of accelerator-based dating techniques. He served from Sublieutenant to Lieutenant in the Royal Canadian Navy from 1944 to 1945. He is the author or co-author of over 230 papers on experimental nuclear physics and accelerator mass spectrometry in various scientific journals.

Chapter 1

An Introduction to the Turin Shroud and to Carbon Dating

Shroud aficionados entering the Cathedral of John the Baptist in Turin are confronted, outside the Royal Chapel, with a full-size, colour photograph of the Turin Shroud. That will have to satisfy their curiosity. The shroud itself is stored, elaborately coffined, on an altar behind a triply locked iron grill in the cathedral's chapel. It is only displayed to the public on special occasions every forty years or so.

The photograph shows an altogether impressive and beautiful stained linen cloth the colour of old ivory, 14' 3" long and 3' 7" wide. It bears the faint front and back imprint of a naked crucified man with hands folded modestly over his genitals. The image depicts all the stigmata of the crucifixion described in the Bible including a large blood stain from the spear wound in the side. The linen weave is a three to one herringbone twill. A seam or tuck divides the main body of the shroud from a 6" side strip of the same weave which runs almost the entire length of the cloth. A backing cloth of basket weave covering the entire back area of the shroud is exposed at both ends of this side strip where pieces of the side strip have either been removed or never existed.

The most notable feature of the shroud is the sixteen patches that were applied symmetrically in pairs to the front of the shroud in 1534, two years after it was damaged in a fire that occurred in the chapel in Chambery, France, where the shroud was stored in a silver chest. Gouts of molten silver burned through the shroud, fortunately outside the image, in a symmetric fashion due to the way in which it was folded in the chest. The shroud was doused with water before the fire damage could spread to the image. This

1

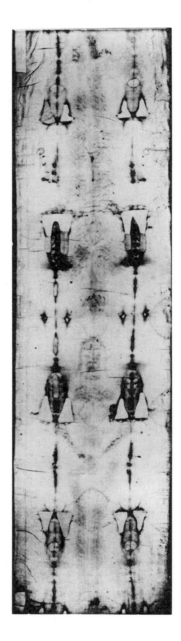

Black and white photograph of the shroud. It is a stained linen cloth of herringbone weave the colour of old ivory, 14' 3" long and 3' 7" wide, bearing the back and front head-to-head image of a crucified man. A very noticeable feature is the 16 triangular patches applied by the Poor Clare nuns in AD 1534 after the fire in AD 1532.

near catastrophe, however, did yield some interesting scientific information. Silver melts at a temperature close to 1800 °F. Because the shroud was folded inside the chest, there had to be a considerable variation of temperature at various points on the image ranging from something near this high temperature to ones approaching normal room values. Yet there was essentially no change in the appearance of the image from one region to another. Since many art pigments volatilize at temperatures well below the melting point of silver, those that could have been used, if it is a painting, are rather limited. The patches were applied by Poor Clare nuns. At the same time, the backing cloth was added to strengthen the shroud linen.

In 1978, I visited Turin to explore the possibility of carbon dating the shroud by a newly developed method requiring very small samples and had the opportunity of seeing it first hand. It was on display in the cathedral and by the time I arrived it had been viewed by close to three million people. The shroud was mounted on a wall at the end of the cathedral behind bullet-proof glass. I found myself strangely moved as I watched the tear stained faces of the people filing by. Could this possibly be the burial cloth of Jesus Christ? Could the faint image be Christ's portrait? As a scientist I thought it exceedingly unlikely but I must say I rather hoped that it was. And another thought occurred to me—why should science intrude itself into the life of this lovely object? It was clearly providing religious inspiration to millions of people. Was this not a case where, although science could provide an answer to its age, it need not?

The shroud's history is thought to date to about the year 1353 when a French knight, Geoffroi I de Charny, claimed ownership. He died in the Battle of Poitiers in 1356 without revealing how it had come into his possession. De Charny was a literate man and the author of the only book on chivalry up to that time written by a layman. He placed the shroud in a church in Lirey, France. It was exhibited in 1357 (the shroud's most secure historic date) shortly after his death amidst considerable controversy. The local bishop felt it was being sold to the faithful as the authentic burial cloth of Christ. He declared that it was a painting and claimed to know the name of the artist. It was exhibited again in 1389 although this time no claim was made that it was Christ's shroud. In the sixty years between 1390 and 1450 details on the Lirey shroud's whereabouts are remarkably vague.

The shroud, or some version of it, eventually passed into the

Present location of the shroud. It is rolled around a wooden cylinder, wrapped in red silk and stored in an ornate silver chest inside a wooden box behind a metal grill with three locks. Located in the Royal Chapel at the rear of Turin's Cathedral of St John the Baptist.

The Cathedral of St John the Baptist.

hands of Margaret, daughter of Geoffroi II de Charny. In 1453 it passed from Margaret to the House of Savoy. In 1502 it was moved to a Savoyard chapel in Chambery, France. In 1578 the seat of the House of Savoy was moved to Turin, Italy and the shroud moved with it. It was used on various ceremonial occasions associated with the Savoy family. The shroud remained a possession of this royal family up to the time of the death in 1983 of the last king of Italy, Umberto II, who willed it to the Vatican.

The shroud remained in Turin except during World War I and II. In 1939 it was moved to the Benedictine monastery of Montevergine in the mountains of southern Italy. Seven years later it was moved back to Turin and continues to reside there despite Turinese paranoia that the pope will someday order it moved to Rome. The archbishop of Turin has been designated by the pope as the custodian of the shroud and the person who makes all major decisions regarding its care and disposition.

A remarkable event in the life of the shroud occurred in 1898 when an eight day exposition took place. An Italian photographer, Secondo Pia, was permitted to take the first ever photographs of the shroud. The equipment of the period required long exposure times under the most brilliant arc lighting during which it is a wonder the image did not pale several shades. Late one night Pia began developing his precious photographic plates. To his wonderment the image slowly emerging on the plates immersed in the developer solution appeared as a positive! To him it was as if he were seeing the face of the Lord!

To this day, the fact that the image on the shroud is a negative is regarded by the faithful as evidence that it was not the creation of an artist but must have been the result of some mysterious burst of radiation occurring at the time of Christ's resurrection. Consistent with this view is the fact that the purported bloodstains are positives, as they should be if the blood from the many wounds depicted on the image soaked into the fabric. Theories have been advanced for a more natural production of a negative image however.

In 1973 the shroud was shown for the first time on Italian television. A forensic expert from Zurich, Switzerland, Max Frei, was permitted to remove pollen samples using sticky tape applied to the shroud surface. Frei, although not a pollen expert, claimed that some of the pollen came from plants indigenous to Palestine. However, professional palynologists tend to be sceptical of claims that pollen is a reliable indicator of provenance. At the same time, two samples were cut by nuns from the hem of the shroud for examination by Professor Gilbert Raes, director of Ghent University's textile laboratory. One sample, about 5 square centimetres in area, was taken from the main body of

the cloth and the other, 3 square centimetres in area, was cut
from the side strip. His examination of the samples under an
electron microscope convinced Raes that there were trace amounts
of Egyptian cotton present in the predominant linen of the shroud.
This could constitute evidence that the shroud was woven on a
loom in the Near East previously used to weave cotton.

He is said to have stored the samples in a stamp box in his
desk and to have brought them out occasionally for the edification
of his guests. A few years later the shroud's custodian, the
then archbishop of Turin, was reminded of their whereabouts
and requested their return. Raes shipped them back by surface
mail. Their safe arrival in Turin, despite the vagaries of the
postal service, has been facetiously suggested as an indication that,
indeed, the shroud must possess some miraculous property. The
scientific endeavours of both Frei and Raes were the precursors
of a long series of subsequent, equally insubstantial, scientific
excursions on the shroud and the nature of its image. The Raes
samples so casually removed for such a pointless examination
were large enough for carbon dating by the newly developed
technique. However, their ill-documented history while in Raes'
keeping was eventually deemed by the science advisor to the
archbishop of Turin to render them unsuitable.

Except for carbon dating, the most recent scientific forays on
the shroud occurred at the conclusion of its latest exposition in
1978. For a period of five days and nights a group of scientists
subjected the shroud to a variety of experiments. Their aim
was to establish that it was, indeed, the burial cloth of Jesus
Christ. They came mainly from US government military and
research establishments and had banded together under the aegis
of the non-profit organization called the Shroud of Turin Research
Project, Inc. (STURP).

They comprised mainly true believers in the shroud's authenticity. They overwhelmed the Turin ecclesiastical authorities and
their scientific advisors with their aerospace technology and their
insistence on military-like secrecy and discipline. Like all the scientific investigations that had gone before, their results were inconclusive and generally of negligible importance despite the time
and money expended. I believed STURP's members to be so convinced it was Christ's shroud that I was determined to prevent
their involvement in its carbon dating, if that were ever to come
about. I feared the most important measurement that could be

made on the shroud would be rendered less credible by their participation. Fortunately in this I was successful.

Possibly not since the days of Galileo has such a curious interaction between science and religion taken place. It culminated in the only measurement that could provide definitive information on a fundamental property of the Turin Shroud, namely the time when the flax used to make the shroud's linen was harvested. Nobody would argue that this measurement had any scientific significance, unlike many others that have been made by the new method called accelerator mass spectrometry (AMS). However, the wide public interest in the shroud and consequently in any scientific technique that could unambiguously establish its age, made it a legitimate object to be tackled by AMS. Funding for the development of AMS came, in general, from US federal sources and that means from tax revenue. If, occasionally, this money is used for projects not of high scientific import but that capture the interest of the general public, that would seem to be proper.

Accelerator mass spectrometry was invented in 1977 at the University of Rochester's Nuclear Structure Research Laboratory located in Rochester, New York by scientists from Rochester led by myself, the University of Toronto in Canada led by Ted Litherland and from a small US corporation called General Ionex led by Ken Purser. It was also conceived independently at the Lawrence Berkeley National Laboratory in California. The first measurements of radiocarbon in nature by this revolutionary new method were carried out at Rochester where the technique was rapidly perfected. It was subsequently widely replicated in other laboratories throughout the world. It quickly became the method of choice for radiocarbon dating of organic matter, especially when the amount of material was limited.

Shortly after the publicity surrounding the invention of this new technique for carbon dating, a letter was received by the Rochester group from a minister in England concerning the Turin Shroud. The letter, dated 24 June 1977, was from the Reverend H David Sox. It was the first time any of us in the Rochester AMS project had ever heard of the Turin Shroud. The question Sox raised was whether this new small-sample technique which he had read about in *Time* magazine could be used to date the shroud. We responded that, indeed, it could but it was a bit too soon to apply so recently developed a technique to such a renowned object. However, his inquiry led, via a complex chain of events, to the actual dating of the shroud cloth by accelerator mass spectrometry almost exactly eleven years later.

Sox, an American Anglican priest, teaches at the American School in London, England. He has long been interested in the Turin Shroud and has authored several books about it. Although in this initial contact Sox gave no indication of his religious denomination or his involvement with the shroud, I learned later that, at the time he wrote to me, he was general secretary of the British Society for the Turin Shroud (BSTS). I had many interactions with Sox in the years that followed and we became quite good friends.

It was he who persuaded me to visit Turin at the time of the October 1978 meeting on the shroud which preceded STURP's five days of testing. He suggested I somehow convey to appropriate Turin authorities the information that a new method of carbon dating had been developed that could establish the shroud's age by using only a negligible amount of cloth. Although I was finally allowed to describe the method of dating briefly to the delegates towards the end of the conference, it had little impact on the shroud custodians.

In the ensuing years, STURP's measurements on the shroud were analysed, discussed at meetings organized by STURP and eventually published in various journals. Their final conclusion was that the shroud was not a painting. This was later disputed by one of the STURP scientists, Walter McCrone, an eminent microscopist from Chicago. He resigned from STURP, having become a *persona non grata* because of his heretical views. Most STURP scientists believed it was most likely Christ's burial cloth. Indeed, in a book he co-authored, Kenneth Stevenson, one of STURP's most militant and religiously zealous members, estimated that the chances against it being the shroud of Jesus were 1 in 83 million. With odds like that it would seem unnecessary to make any further measurements. His calculations, however, made no sense to me.

It is well known to scientists that one can sometimes obtain a desired scientific result by subconscious manipulation of the technique or the data. It is a human flaw that must be carefully guarded against. It is most easily circumvented by not having preconceived notions of what the answer should be. A belief that the shroud was the genuine article was the stuff of which STURP was made and I am happy to say that, in the end, they played no role in its carbon dating.

In 1984, the British Museum coordinated an interlaboratory

test of the ability of several laboratories to measure the age of cloth. It included laboratories using the AMS method developed at Rochester and those using a conventional counting technique that could also deal with small samples, developed at Brookhaven National Laboratory (BNL). By now six laboratories, four using AMS and two using small counters, had expressed a willingness to date the shroud if the Turin authorities wished to have it done. The results clearly indicated that these six laboratories were competent to do so.

At the *12th International Radiocarbon Conference* held in the spring of 1985 in Trondheim, Norway, the results of the interlaboratory comparison were presented. I arranged an informal meeting of representatives of the six laboratories and the British Museum to discuss what should be done next. It was agreed that, to ease the communication gap that existed between science and religion (at least as exemplified by the Roman Catholic Church in Turin), it would be advantageous to establish contact with the Pontifical Academy of Sciences in Rome. This group of some 75 eminent scientists in all fields and of every religious persuasion, including agnostics, was chaired by Professor Carlos Chagas, a medical doctor and biochemist from Brazil. It was formed over fifty years ago to advise the pope on matters of science affecting the church.

In the weeks that followed, I managed to contact Professor Chagas through the good offices of a theoretical astrophysicist, Dr Vittorio Canuto, a colleague of Chagas, who works in the Goddard Space Science Center of the National Aeronautics and Space Administration (NASA) in New York City. This led to the organization of a workshop on the Turin Shroud held in Turin in the fall of 1986 under the joint auspices of the Pontifical Academy of Sciences and the Archdiocese of Turin and chaired by Chagas. This workshop produced a protocol that was to involve seven laboratories in carrying out the carbon dating measurements on the shroud (a fifth AMS laboratory located in France had joined the group since the Trondheim meeting). The protocol was subsequently rejected by Turin ecclesiastical authorities. In particular, the seven laboratories were reduced to three. Inexplicably, Rochester was not one of them.

Despite attempts on my part to persuade the laboratories to stick together and to insist on following the original protocol, the three finally agreed to proceed. All three employed the AMS technique and so this revolutionary new carbon dating method

with which I had been so closely associated for almost ten years
was about to be put to its most stringent public test.

The first measurement (at which I was present) was made
at the University of Arizona on 6 May 1988 using their
tandem accelerator facility specifically designed for carbon dating.
The other two laboratories, in Oxford, England and Zurich,
Switzerland, a few weeks later independently confirmed the result.

This book, written after the result of the carbon dating had been
announced and after some time had elapsed for it to be digested,
gives a brief historical account of the development of the scientific
method that made that dating possible and of the fascinating
and often exasperating interaction between science and religion,
and within the scientific community itself, that culminated in the
final answer. It is also about integrity in science, in personal
relationships and in religion.

The Turin Shroud would never have been carbon dated without
the invention of accelerator mass spectrometry. As stated earlier
in this chapter, the invention and development of AMS occurred
at the Nuclear Structure Research Laboratory at the University of
Rochester—a laboratory I founded and which began operating in
1966. It houses a tandem Van de Graaff accelerator designed for
research in nuclear physics and is one of the top such university
laboratories in the USA. Until 1977 it had made only a few modest
forays into fields other than nuclear physics. Since then a fifth of
the accelerator's running time has been devoted to AMS.

How does a tandem Van de Graaff accelerator work and what
makes it valuable for AMS? Briefly, a tandem starts by converting
neutral atoms to negative ions by the addition of an extra electron
to the atom in an ion source at the entrance to the accelerator.
Details on how this happens need not concern us but, to many
people, the phrase 'negative ions' conjures up the possibilities of
libido enhancement. It is claimed that powerful concentrations
of these erotic negative ions are found at high altitudes and in
the vicinity of waterfalls—like those in my home town of Niagara
Falls. Could this be why it has been considered the honeymoon
capital? In an advice column in a local newspaper a retired
Canadian specialist in internal medicine suggested to an elderly
male correspondent that, to improve his sex drive, Niagara might
be worth a gamble. As I reach retirement age and beyond, such
advice makes the thought of returning to my home town grow
increasingly tempting.

Unfortunately, in a tandem accelerator the negative ions are securely confined and so their alleged salubrious effects are effectively negated. In any event, after their production in the tandem's ion source, these singly charged negative ions are accelerated toward a positively charged high-voltage terminal in the centre of the tandem. There several electrons are removed in the terminal stripper and the now multiply charged positive ions are accelerated away from the terminal back again to ground. This high-energy beam of the desired projectile nuclei then bombards a thin target of the desired target nuclei, inducing nuclear reactions. It is a powerful and versatile machine for exploring the intricacies of the atomic nucleus and, as we shall see, for AMS.

To carbon date organic material such as the cloth of the Turin Shroud, it is necessary to measure the relative amounts of various isotopes of carbon in the material. Isotopes of chemical elements are atoms that have the same number of protons in their nucleus (and thus the same number of electrons circling the nucleus) but different numbers of neutrons. Carbon has six protons and six electrons and the latter determine its chemical properties. The stable isotopes of carbon have either six or seven neutrons. These are called carbon-12 (6 protons plus 6 neutrons) and carbon-13 (6 protons and 7 neutrons) respectively. There is a radioactive isotope of carbon with a relatively long lifetime that has 6 protons and 8 neutrons (carbon-14). The relative amount of carbon-14 compared to stable carbon in an organic material is a measure of when that organic material died.

Carbon dating was invented by Willard Libby in the 1940s and earned him the Nobel Prize in chemistry in 1960. Radioactive carbon-14 is produced in the atmosphere by cosmic rays and, along with the stable isotopes of carbon, it combines with oxygen to form carbon dioxide gas. All living organisms ingest carbon dioxide. As long as plants or animals are alive the ratio of carbon-14 to stable carbon in their carbon component is about one part in a trillion. This ratio represents the equilibrium between the production of carbon-14 in the atmosphere and its decay. At the time of death of the organic matter the ratio begins to decrease. If death occurred 5730 years ago (the half-life of carbon-14) the ratio is only half a part in a trillion. A measurement of the ratio determines the time of death with considerable precision.

For many years it had been the dream of carbon daters to employ a mass spectrometer that would directly detect radioactive carbon-14 at the exceedingly low concentration at which it occurs in living organisms by its mass alone rather than waiting for it to decay as Libby had done. Such a method would reduce the amount of material required to make a dating by several orders of magnitude over the decay counting technique then in use.

All attempts had failed because carbon-14 weighs the same as the abundant and ubiquitous stable element nitrogen-14 to a high accuracy. Nuclear physicists familiar with tandem accelerators knew that when a beam of nitrogen ions was required, the tandem's ion source had to be tuned to a negative molecule of nitrogen, such as nitric hydride, rather than the nitrogen atom itself. It was assumed that negative nitrogen ions by themselves were relatively unstable. If they were sufficiently unstable, the tandem's use of negative ions would instantly solve that most serious interference problem. Traditional mass spectrometers start with positive ions—neutral atoms with one electron removed—and it is as easy to rip one electron from neutral nitrogen as it is from carbon. Then why not use negative ions in conventional mass spectrometers? The answer is that there is another serious source of mass 14 interference—molecular hydrides of stable carbon atoms—and these do form stable negative ions. Such molecules are destroyed if enough electrons are removed in the terminal stripping process.

The questions were: just how unstable was the negative nitrogen ion and how many electrons needed to be removed from a carbon hydride molecule to cause it to blow apart? The best way to answer such questions was to make appropriate measurements on a tandem Van de Graaff accelerator. I volunteered the use of the Rochester tandem accelerator for this purpose.

The first AMS experiments on the Rochester tandem took place from 14–20 May 1977. The three senior scientists involved were Ken Purser of General Ionex Corporation, Ted Litherland of the University of Toronto and me. The main aims of the experiments were to determine whether the negative ion of nitrogen was sufficiently stable to reach the tandem's high-voltage terminal (it was not) and how many electrons had to be removed from neutral molecular hydrides of carbon to cause them to completely dissociate (three or more). We immediately knew that we would be successful in our quest to directly detect carbon-14 in natural biological matter.

I purchased a bag of hardwood barbecue charcoal as an example of contemporary carbon since it came from trees that had been felled fairly recently. We inserted a milligram or so in our ion source along with a second sample of graphite as a measure of the background. Graphite is a form of carbon that derives from oil and oil deposits are millions of years old. Any carbon-14 with its half life of 5730 years in graphite derived from oil would have decayed to nothing long ago. On 18 May 1977 the two samples were run sequentially. The charcoal sample gave over a thousand times more carbon-14 counts than did the graphite! Jubilation reigned in the laboratory. We had done it! Now the amount of material needed for carbon dating was so small (at least a thousand times smaller than had been required up till then) that it would be possible to date even the most precious artifact. It was one of those sweet, instantly recognizable triumphs that occur all too infrequently in science. Nothing quite like it had ever occurred to me before.

Once we had obtained this important result, we considered how best to publish it quickly. By tradition, first publication in the news media is unacceptable. On the other hand most scientific journals take so long to publish papers that priority, which is important in science, often gets lost. However, a conference on accelerators was to be held in Strasbourg in a few days. Ken Purser had already been invited to give a talk on unrelated apparatus his company was building for a laboratory in Japan. It seemed remarkably timely for him to substitute a paper co-authored by all of us in the group on the AMS work we had just completed. I phoned the conference organizer, who was a friend of mine, to request that Ken be allowed to substitute a paper on the use of electrostatic accelerators as ultrasensitive mass spectrometers. I told him that it involved a new method of carbon dating that might well revolutionize the field. Ken did deliver the paper and it was to an appropriate scientific audience—a prerequisite to any press release.

Because the proceedings of a conference like the one in Strasbourg reach a very specialized scientific community, we discussed which journal covering a much broader scientific spectrum would be most appropriate for us to submit a more carefully crafted paper. The journal *Science* in the USA and *Nature* in Britain are both prestigious and both cater to a readership covering a wide range of science. We settled on *Science* and began work on the paper.

Our second run took place during early June 1977. We repeated our previous measurements with even better results. At this stage we decided to approach the University's Public Relations Office with the idea of issuing a press release. It was sent to the usual news agencies on 8 June 1977. It stressed the fact that the new technique, reported at a scientific meeting in Strasbourg, represented an important breakthrough not only by reducing the sample size by several orders of magnitude but also in allowing much more ancient artifacts to be dated. It turned out that the former was the more important feature of the method. The release was picked up by a number of news organizations including the *New York Times* and an article written by one of their senior science reporters, Boyce Rensberger, was published in the 9th June edition. I read it on a visit to the Brookhaven National Laboratory on Long Island, New York. Shortly after the *New York Times* article appeared, I got a call from an associate editor of *Time* magazine, who interviewed me for a story on this new dating technique.

On 24 June the issue of *Time* magazine dated 27 June 1977, containing the article on the new carbon dating method, appeared on the news stands. It followed an article on Werner von Braun who had died the week before. Von Braun was the famous German rocket scientist whose buzz bombs and V2 rockets had so plagued London in World War II. After the war, his apolitical proclivity brought him to the States where he became the darling of the US military establishment. The carbon-14 article was titled in the usual brash *Time* style 'The New Dating Game'.

It was during another AMS run in May 1978 that Litherland, Purser and I agreed that we should explore the possibility of getting involved in dating the Turin Shroud. As mentioned above we had learned of it in June 1977 in a letter from Reverend David Sox. I had already privately decided that it would be too good an opportunity to miss. It would be a highly public demonstration of the power of carbon dating by AMS. Besides, I was becoming increasingly curious about the shroud and the, to me, remote possibility that it actually was Christ's burial cloth. I was pleased that they concurred. I had not the slightest inkling how byzantine the project would turn out to be nor that it would be ten years almost to the day that we agreed to be involved in the Turin Shroud adventure that it was first dated by AMS at Arizona. Some time in late July 1978, we sent a paper to Turin describing how we would go about dating the shroud and offering to do so if the authorities there wished.

Chapter 2

The 1978 International Congress on the Shroud

This was the letter from David Sox, dated London, 24 June 1977, that introduced us to the Turin Shroud:

"Dear Prof. Gove,

I write this in strictest confidence. You may know about the Turin Shroud or not. A great deal of very quiet diplomacy has been going on in hopes to have it carbon dated using small samples which have already been removed for other purposes. I have been involved in this process and Dr. Walter McCrone has been a valuable contact. Under no circumstances would I want him to know that I have approached you but there is a great deal of discussion as to whether his methods (or access to new methods) could bring about the best test. He is a marvellous person and I am very devoted to him but my experience with Prof. Apers of Belgium led me to question whether such a test was possible. The enclosed articles will 'spell out' some of this. I do not have a detailed analysis of his process but you will be able to check on Prof. Apers' analysis of it. I know this is a rather strange letter 'out of the blue' but would appreciate your comments on the enclosed and am certain you can keep this in a confidential manner—that is extremely important. I heard about your new testing from a friend who works for *Time* magazine. My best wishes—

(The Revd) H David Sox"

Enclosed with the letter was an article from the 25 March 1977 issue of the *Los Angeles Times* headlined 'Christian Relic under Gaze of Science'. It reported on a meeting held in Albuquerque, New Mexico that had ended the previous day. The purpose of the

73 CHATSWORTH COURT,
PEMBROKE ROAD,
LONDON. W. 8.

24 June 1977.

Dear Prof. Gove,

I write this in strictest confidence. You may know about the
Turin Shroud or not. A great deal
of very quiet diplomacy has been
going on in hopes to have it
carbon dated using small samples
which have already been removed
from it for other purposes. ...

... I know this is a
rather strange letter "out of
the blue" but would appreciate
your comments on the enclosed
and am certain you can
keep this in a confidential
manner — that is extremely
important. I heard about your
new testing from a friend who
works for Time magazine. My
best wishes —

(The Real) H. David Sox

My involvement with the shroud began as a result of this letter from the Reverend H David Sox. He turned out to be the Secretary General of the British Turin Shroud Society. (Courtesy of H David Sox.)

meeting was to plan what became STURP's five day and night investigation of the shroud. The participants hoped this would take place in October 1978 at the end of a six week exposition of the shroud in Turin, Italy.

Among those attending the meeting in Albuquerque were scientists from Pasadena's Jet Propulsion Laboratory, the Air Force Academy (Eric Jumper and John Jackson) and nearby Los Alamos Scientific Laboratory (Ray Rogers and Robert Dinegar); a scientist who claimed to have proved the Vinland Map a fake (Walter McCrone); an eminent British New Testament scholar–theologian (Bishop John A T Robinson of Trinity College, Cambridge); Catholic priests from New York (Reverend Peter Rinaldi, S.D.B.), from the Vatican, and four long-haired members of the communal Christ Brotherhood who just happened to drop by.

They discussed, among other things, the possibility of a radiocarbon dating that McCrone was warmly advocating although none of the participants had any expertise in that field. McCrone thought that half a square inch of cloth would be all that was needed and this amount of cloth had already been removed. Rinaldi said he didn't think the Turin authorities would be against carbon dating but they would have to be convinced that such a small amount of cloth would suffice. Rinaldi was to play a key role in helping arrange the STURP incursion on the shroud in October 1978.

Other enclosures in Sox' letter were a document from McCrone concerning the possibility of carbon dating the shroud using small samples by a new method he suggested and an analysis of the method by Professor D Apers of the University of Louvain's Laboratory of Inorganic and Nuclear Chemistry, concluding it would not work. McCrone's proposal was addressed to Rinaldi who, at that time, was pastor of the Corpus Christi Church in Port Chester, New York and vice president of the Holy Shroud Guild. Rinaldi was an altar boy in the Cathedral of John the Baptist in Turin where the shroud was stored—he had grown up surrounded by its aura. As soon as I met him, some time later, he became to me 'Mr Shroud'.

Meyer Rubin, head of the radiocarbon laboratory of the US Geological Survey in Reston, Virginia, was present during our third AMS run that began in late June. He had contacted us the day after the article on our work had appeared in the *New York Times*. We had invited him to visit Rochester in order that we might benefit from his expertise as a carbon dater. I discussed Sox' letter with him and the others in the AMS group during this run. Discussing the matter with Rubin was no violation of Sox' request for confidentiality because I felt Rubin should collaborate on the project if we decided to take it on. It turned out that none of us had ever heard of the Turin Shroud. In my reply to Sox dated 6 July 1977, which I signed on behalf of Litherland, Purser and myself, I said that we were not yet in a position to accept samples for dating. However, the method would ultimately permit objects like the Turin Shroud to be dated using very much smaller samples than heretofore possible, including the method proposed by McCrone.

McCrone had suggested that one of the shroud samples removed in 1973 for examination by the Belgian textile expert, Professor Gilbert Raes, could be placed on a photographic plate. Then, after some time, the plate could be developed and the tracks made by the electrons emitted when carbon-14 decayed could be counted and the sample dated. The idea is scientifically nonsensical on several counts as Apers, more tactfully, pointed out. In my letter to Sox I merely suggested it probably was no match for our AMS method.

By coincidence, on the same day I replied to Sox I got a call from Walter McCrone. We had never met and obviously he had no knowledge of the Sox letter. He told me, in a rather secretive manner (the shroud seems to bring out conspiratorial traits in people), that he had an important piece of cloth whose age he was eager to know. In particular he said he would like to know whether it was 2000 years old. Presumably he thought I couldn't guess to what cloth he was alluding. Had he called before we received Sox' letter I would, indeed, not have known. I had begun to suspect that Sox and McCrone were cooking up some scheme to obtain Raes' shroud samples. I told McCrone he should write to me and he said he would.

He did so on 11 July 1977. He described the Walter C McCrone Associates, Inc. laboratory in Chicago. It was devoted principally to applied microanalytical research. He included some reprints of recent work he and his associates had been doing on the Vinland Map.

That map depicted land in the north-west Atlantic, west of Greenland, that could correspond to the Vinland of the Norse Sagas. There was reason to believe the map was pre-Columbian and thus constituted evidence that America was first discovered by the Norse. In 1965, President Lyndon B Johnson had proclaimed 9 October as Lief Ericson Day. The owners of the Vinland Map, Yale University, first revealed its existence on 11 October 1965. They claimed it was a mere coincidence that this was the day before Columbus Day. It just happened, they said, to be the closest convenient date to Lief Ericson Day!

McCrone Associates were later commissioned by Yale to investigate the map's authenticity. McCrone found that some of the white ink on the map contained the pigment anatase. Anatase (titanium oxide) was not used as a white pigment before 1917. Well aware of the publicity value, he released information to the press that the map was a fake. The news was greeted with relief and joy in Italy and in the Italian-American community but officials at Yale were not amused. More recently, doubts have been raised concerning McCrone's conclusions and there is increasing evidence that the Vinland Map predates Columbus' discovery of America.

In his letter, McCrone remained coy about the source of the cloth he said he was interested in dating. He stated, however, that it was a very important sample believed to be of the order of 2000 years old. He thought it would become available to him in the very near future. He went on to say there were two samples weighing 60 milligrams and 110 milligrams respectively which must be dated to dependably distinguish between ages of 2000 and 700 years and, for some unstated reason, he needed the results before the end of the year. They were, as I already knew, the Raes samples and I suppose he wanted the results before the next shroud exposition scheduled for early fall of the following year. He did not, needless to say, reveal how he planned to obtain the samples. He also mentioned he would be exploring Muller's cyclotron method as well as some refinements in the classical carbon dating procedures in progress at Brookhaven (that was the first we had heard of this).

I did not use our standard form letter of regret that we were not yet ready—the letter sent to others requesting we date some artifact or other—but sent him a copy of our Strasbourg paper and the paper we had sent to *Science*. I said we were not yet in a position to date his important samples but, conceivably, might be in six months or so. I did not reveal that I knew what the samples were. I invited him to visit our laboratory. I sometimes think that McCrone dreamed of becoming history's greatest iconoclast. Having, in his view, demolished the authenticity of the Vinland Map he saw the chance to do the same to the Turin Shroud!

When a series of tests were carried out on the shroud in the fall of 1978, McCrone determined that there were traces of iron oxide powder on the shroud image. He immediately announced that he "had some good news and some bad news. The bad news is that the shroud is a fake. The good news is that no one is going to believe me." It remained, however, for others to settle the question and to do so with somewhat greater objectivity and with a great deal more credibility.

On 11 May 1978 I phoned the Reverend H David Sox at his home in London to find out what the latest interest was in carbon dating the shroud. He told me that the samples removed from the shroud in 1973 for examination by Professor Gilbert Raes had been in the hands of the Turin ecclesiastical authorities ever since Raes had mailed them back. He and McCrone had been working together to arrange for these samples to be made available for carbon dating. However, Turin had become disillusioned with McCrone and had publicly announced that they would not make these samples available for dating. This information was considered so startling that it was carried as a major news item in the *Times* of London.

I said we might be ready to date the shroud by AMS in a month or so but I raised a question that had been troubling me for some time. Why should the shroud be dated? Would it not profoundly disappoint many people whose religious faith was buttressed by their belief that this was the burial cloth of Christ if, for example, it actually turned out to be only 800 years old? Sox said that it was important to discover the truth. Even King Umberto, technically its owner, who was then living in exile in Portugal, was keen to have the carbon dating measurement made. Sox believed that the archbishop of Turin would eventually agree.

Sox then told me that a special exposition of the shroud would be held in Turin between 27 August and 7 October 1978. After that, it would be returned to its silver casket and might not be available again for any purpose for some considerable time. He said my phone call was particularly timely and he would make further inquiries and get back to me. I wrote Sox on 12th May expressing our possible interest in dating the shroud.

Around mid-May, I received a letter from Sox suggesting that it would be useful for me to attend the October congress on the shroud to be held in Turin. He pointed out that it would be an excellent opportunity to meet not only the Turin authorities directly associated with the shroud but also the people comprising the Shroud of Turin Research Project. STURP would be conducting tests on the shroud following the congress and the shroud's six week exposition. It was something I very much wanted to do but did not know how I would find the money for the travel expenses. I could hardly allocate them from the National Science Foundation's operating grant to the laboratory. A few days later I got another letter from Sox saying that he was planning to visit his parents in North Carolina sometime around the end of June and he would also like to visit Rochester. Did this mean he was an American? I had assumed all along that he was a minister with an American accent. In fact he is an American—an Anglican—who teaches at the American School in London. At the end of May I wrote to him inviting him to visit during the week of our 9th run, 26 June–2 July, and to stay at my house.

The same day I wrote to Sox inviting him to visit Rochester, I received a call from Kathy Smalley of the Canadian Broadcasting Corporation (CBC) in Toronto. She was the producer of a CBC television programme called *Man Alive* which dealt with a variety of interesting and somewhat off-beat but true-life subjects. She was planning a programme on the Turin Shroud. She said she would like to bring a CBC television crew to Rochester during our next run. She had learned about our interest in dating the shroud from Ted Litherland and Roelf Beukens at the University of Toronto. I was rather diffident at first but, since her contact was through our Canadian collaborators, I suggested she come on 28 June.

Smalley called again the next day and said the CBC camera crew would arrive on the date suggested. She had talked to both Meyer Rubin and Lucy McCrone and said both of them would be in Rochester on that day. Meyer really captivated her as he is wont to do. I wondered if he had used the hoary line he once told me he found so effective with women: "My mother's name is Kathy". Lucy McCrone, Walter McCrone's wife, had attended our April AMS meeting. Kathy planned to bring a full-size replica of the shroud to Rochester and had in mind televising Lucy examining the shroud under a microscope. I thought to myself 'The hell she will'. Fortunately Lucy was unable to make it.

That day I also got a call from McCrone. In his opening gambit he gratuitously donned the mantle of leadership in getting the shroud dated despite lacking expertise in the field of carbon dating. He opined that two laboratories would have to date the shroud independently, for example Rochester and Berkeley. I gently demurred by stating that Berkeley was not yet ready, if they ever would be, which I greatly doubted. I suggested that Oxford might be ready before too long. Arriving at the notion that it should be done independently by two or preferably more laboratories did not take a giant leap of intellect. McCrone said that, of course, the former King Umberto should be informed of the results and would have to decide whether or not they should be revealed to the public. Was this the same McCrone who contacted the press with the news that the Vinland Map was a fake allegedly without first getting permission from his client, Yale University, because he had found traces of titanium oxide in the white pigment? I stated flatly that was not acceptable. If the shroud were ever carbon dated, the results would be published whether or not anyone objected.

He mentioned that the two pieces removed in 1973 came from the hem of the shroud and thus might be of more recent vintage. He proposed to get some threads closer to the image. He did not reveal how he planned to do this. We discussed how much material would be needed and I said enough to provide from one to ten milligrams of carbon. He thought that would be no problem. When he suggested that the results of the measurements should be available before the shroud was returned to its cask on 8 October following its six week exposition, I warned him that we would not accept any deadline for producing a result. I told him that David Sox would be visiting us the week of 27 June. He said his wife would also be coming then and that she had been very impressed with our April AMS meeting. He wondered when the proceedings would appear and I said in two weeks.

On 26 June we began another series of carbon-14 measurements. We wanted to improve the accuracy and reproducibility with which we could carry out measurements on the natural organic samples supplied by Meyer Rubin. We wanted to refine our technique to the point where it was truly competitive in this respect with the standard decay counting method.

On the third day of the measurements, Lee Krenis came by. She was a member of the university's public relations department. Among other duties, she covered the science and medicine part of the research going on at the university. She had shown a great interest in the possibility that we might date the Turin Shroud. One of the problems public information people have with many of the important stories in the physical sciences is their inherent complexity and the difficulty of explaining them in a comprehensible and interesting way to the general public. A story on dating the shroud would clearly be a real winner. She had phoned me early in June concerning the Canadian Broadcasting Corporation visit, offering to help in any way she could by shepherding them around, taking them to lunch or dinner, etc.

David Sox also arrived that day. It was the first time I had met him and it began a long personal relationship over the ensuing years. I suppose at that time he was in his late forties, very dapper whether in clerical or civilian garb and he favoured the latter, invariably wearing modish clothes—a man of considerable charm and humour. He can occasionally be mildly effeminate with a slightly shrill laugh. He has a tendency to be a name dropper in conversations. On the other hand he does know many important people and he knew a lot about the Turin Shroud and the people associated with it.

As an example of his acquaintances, my wife had spotted a brief article in the 17 July 1977 edition of the *Democrat and Chronicle*, the Rochester evening newspaper, that said the Reverend David Sox had given the funeral oration at the burial of the widow of King Carol of Romania. She had died in her villa in Estoril, Portugal on 30 June. She was identified as Madame Magda Lupescu who had a widely publicized love affair with King Carol in the 1930s. He later married her. Her casket was placed on a dais beside his in the Royal Pantheon in Lisbon where—so the news report said—Portuguese royalty are entombed. I had written to Sox on 12 May 1978 following my phone call to him deploring the decision by Turin not to make the Raes samples available for carbon dating. In my letter I mentioned this brief newspaper article and asked if I was correct in assuming he had presided at the funeral. He wrote me that, indeed, he had officiated at Princess Elena's (Madame Lupescu's) funeral last summer. He was the priest in charge of the Anglican church in Estoril for two months and that was his first duty. He said this was near where King Umberto lives, so he had gotten to know him better then. I was impressed.

Kathy Smalley arrived with the CBC crew in tow. There were six altogether including Roy Bonisteel who hosts the *Man Alive* programme, serves as the narrator and interviews the people involved in whatever subject is being covered. The four others comprised a cameraman, a script girl (the CBC was still using that appellation) and a couple of electrical technicians. She also brought along a full-size replica of the shroud. Sox was interviewed on camera standing at the high-energy end of the tandem. Meyer Rubin and I and several others involved in the experiment were videotaped peering owlishly at bits and pieces of the accelerator complex or uttering profound comments to Bonisteel's promptings. It was all great fun and formed an important part of the final programme Smalley produced for the

CBC. It was shown on Canadian television some time after the October congress on the shroud held in Turin.

I am sure the Nuclear Structure Research Laboratory tandem crew and other laboratory people not directly involved with the AMS operation got quite a thrill out of the attention that was being paid to the laboratory. That sort of thing is good for morale. Local TV news people had come to the laboratory on several occasions in the past since our AMS programme began just over a year ago, but never a major production like this.

In my discussions with Sox, he said it was important for me to attend this Turin congress on 7 and 8 October 1978. It would be a scientific meeting; I would meet Father Peter Rinaldi and other local people connected with the shroud. Among the members of Turin Shroud societies who David thought might have a say as to who should be involved in dating the shroud, Rinaldi was 'numero uno' (his phrase) and, in all modesty, he said that he was number two. As we were both to later learn, Luigi Gonella in his capacity as science advisor to the archbishop of Turin on matters of the shroud was 'numero uno' and all the other numbers as well!

He said that 9 October 1978 would be the date on which the nuns would remove sample threads from the shroud if samples were permitted for carbon dating. He made it sound as if there was a serious possibility this would happen. That seemed unlikely to me. He went on to say that when these threads were removed, it would be important for someone from our group to be present in order to have a say in the selection of sample sites and size. Walter McCrone would be there and wanted to play a major role in the dating operation.

That evening Lee Krenis and I wined and dined David Sox and Kathy Smalley at the University of Rochester's Faculty Club. It was quite an entertaining evening with Sox recounting the many adventures he had already had with the shroud, McCrone and other characters associated with the shroud. A few days later Lee Krenis wrote me a short note saying how much she had enjoyed the shroud discussions, the CBC filming and the fascinating dinner with Sox and Smalley. She said she was now hopelessly captivated by the whole business. She warned me that the CBC people were extremely interested in the financial aspects of the shroud business, as was apparent from some of Kathy Smalley's questions at dinner. In a postscript she added "the next day Roy what's his name, the interviewer with the fangs, asked me specifically about Ken Purser in the context of being curious about those who might be in a position to profit financially from the shroud. As you know, it is easy to create an atmosphere of suspicion and intrigue even when there is nothing going on; 60 Minutes does it all the time. I hope CBC's intentions are as genuine and above board as they seem." Fortunately they were, because there were no innuendos about such matters in the *Man Alive* programme they finally produced.

In the 21 July 1978 issue of *Science,* an article appeared in the News and Comments section titled 'The Mystery of the Shroud of Turin Challenges 20th-Century Science'. It stated that, in celebration of its 400th anniversary in Turin, the shroud would be on public view from 27 August through 8 October. An American team, coordinated by physicist John J Jackson of the US Air Force Academy in Colorado, might have access to the shroud after that time.

The article described the shroud, its known history, previous investigations of its remarkable characteristics and plans the Jackson team had for further tests. It stated that Walter McCrone, whom it described as one of the few scientists to become involved in the Shroud study independently of the John Jackson connection, wanted to vacuum the shroud to collect particles for examination back in his Chicago laboratory. He eventually hoped to receive permission to apply carbon dating to the shroud.

The article went on to say that, although current carbon dating techniques required too large a sample, McCrone anticipated that the availability of new technology employing 'linear accelerators' could accurately date a one centimetre length of a single thread. McCrone was a bit optimistic on the sample size required but, at least, he realized the accelerator would have to be linear (like our tandem) rather than circular (like the cyclotron at the Lawrence Berkeley National Laboratory in California—our AMS competitor).

This article in *Science* was mainly lauded but also excoriated in subsequent letters to the editor. On the whole, however, it lent a certain aura of respectability to scientific explorations of the shroud. It was a matter that worried me from time to time. As a scientist, can one justify spending any of one's professional time on a religious artifact? My main justification was that dating the shroud captured the public imagination and it would be a tremendous boost for this new and publicly unknown technique. Other than this, it would be difficult to argue that dating the shroud served any scientific purpose.

David Sox phoned the laboratory in early August asking

whether we intended to submit a paper to the Turin Shroud congress in October. I had felt hesitant about requesting travel funds to attend the meeting from the NSF but I managed to persuade Ken Purser to have General Ionex provide the money. With that settled, on 1 September 1978 we submitted a paper to the congress on how we would propose to carbon date the shroud by the AMS method.

In the paper, titled 'A Method of Dating the Shroud of Turin', we suggested that an amount of shroud cloth sufficient to produce of the order of 10 milligrams of carbon would be ideal. Exactly what this would correspond to in terms of the length of thread or area of cloth depended on the cloth or thread density and the cloth-to-carbon conversion ratio which were, to us, imperfectly known. From some estimates we had, it seemed this might require one meter of weft thread and three times as much of warp thread.

McCrone also submitted a paper to the congress in which he used a more accurate estimate of the cloth-to-carbon conversion ratio that reduced the thread lengths required by a factor of almost three. In either case, the amount of material required would be a thousand times smaller than required by the conventional decay counting method. Both papers eventually appeared in the congress proceedings accompanied by translations into Italian.

On 25 September 1978 I left on the trip that ultimately took me to Turin. I visited Oxford first and met with Robert Hedges who would be the person directly in charge of Hall's AMS facility in his archaeometry laboratory. Hall at that time had submitted a proposal for the necessary funding for his AMS carbon dating facility which Ken Purser hoped would be based on one of his Tandetrons. I told Hedges I was on my way to Turin to participate in a scientific congress on the shroud with a view to presenting a proposal for carbon dating it. He let me know, in no uncertain terms, that he considered such an involvement with a religious object like the shroud to be quite unprofessional and that, if he had any say in the matter, Oxford would never be involved in such an enterprise. Hedges is a very competent scientist but rather conservative in a typical British way. I must say I was a bit surprised at his attitude none the less. When I mentioned Hedges' comment later to his boss, Teddy Hall, Hall remarked that he had no such similar reservations, a fact he amply demonstrated in the years to come.

The next day, I gave a seminar at the Oxford Nuclear Physics Laboratory located next door to Hall's laboratory. The former is a building housing a large home-built tandem designed by Dick Hyder as well as a smaller tandem manufactured by the same company that built ours, High Voltage Engineering Corporation. The director of the laboratory was Ken Allen. I had known and had close associations with both Hyder and Allen for many years and I was to spend a sabbatical there in 1983–84. Hyder told me that he was refereeing Muller's proposal to upgrade the Berkeley cyclotron for AMS. I had suggested him for that to the NSF's Bill Rodney. I was curious to know what Hyder thought of the proposal but did not ask.

Hall thanked me for the material we had provided him on the actual carbon dating we had carried out, which he had used as an addendum to his proposal to the British funding authorities for an Oxford AMS facility. This was to convince them that the technique worked as advertised and Hall said it had been very helpful.

That afternoon, I travelled by train and bus to London's Heathrow Airport and took Air Alsace to Strasbourg. I gave a colloquium at the Centre for Nuclear Research (CNR) in Strasbourg where I had spent a sabbatical in 1971–72. The next day, I left Strasbourg by train for Munich and a day later I visited the tandem laboratory in nearby Garching. It was jointly operated by the University of Munich and the Technical University of Munich. I gave another talk on the work we had been doing on the detection of carbon-14 and chlorine-36 using a tandem identical to the one operating at Garching. The Munich tandem would enter the AMS field before long.

I left by train the next day for my next stop, Heidelberg, and drove to the Max Planck Institute for Nuclear Physics that housed their tandem accelerator. While I was chatting with one of the members of the laboratory, he got a phone call from Hans Suess to say that both he and Professor Otto Haxel would be at my talk tomorrow. I felt honoured. Haxel, Jensen and Suess had played a leading role some years ago in formulating the so-called nuclear shell model and Suess had, more recently, attained additional prominence in contributions he had made in the field of radiocarbon dating. My talk was at 2 pm and there was a very large audience which included these two famous professors. I remember writing to Suess on a later occasion and spelling his name Seuss to which he took great umbrage—"I do not write children's books!"

I returned to Strasbourg by train and from there flew to

Rome on 5 October 1978. When I arrived in Rome the first person I spotted was Roy Bonisteel of the Canadian Broadcasting Corporation, who was taking the same flight as I to Turin. Also in the Rome airport were a US medical doctor and Father McGuire, a member of the Holy Shroud Guild, who were also going to Turin to attend the congress. McGuire was visibly unimpressed with my description of our new carbon dating method. He mistook the 20 centimetres of thread I said we would need to date it for 20 metres and assumed we would need a whole clothesline worth.

When I arrived in Turin I checked into my hotel and then walked to the Cathedral of John the Baptist around 4 pm. The shroud normally reposes in a cask in the Royal Chapel connected to the cathedral but it was now on display to the public in the cathedral itself. During my walk I noticed that every important building was guarded by armed and patently nervous uniformed young men, presumably because Turin was the headquarters of the notorious and then very active Red Brigade. In the square outside the cathedral there were mobs of people lined up to see the shroud as there had been ever since it went on display over two months ago. I walked back and had dinner in a restaurant near my hotel where I met several Americans who were all animatedly discussing the shroud.

About 9 pm I went back to the cathedral, entered and sat near the back. I saw the shroud for the first time quite a distance down at the other end of the cathedral. It was mounted behind bullet-proof glass above the cathedral altar. It was too far away to get any real impression of its beauty and of the wealth of crucifixion details it contained. Crowds of people were still entering the cathedral, walking slowly down an aisle on one side, crossing a wooden bridge at the far end below the shroud and some twenty feet from it, and then leaving by an aisle on the other side to exit the cathedral.

The next morning, on 6 October, I changed my hotel to the Venezia, that Sox had suggested in the first place. It was within a stone's throw of the cathedral and would be where the British delegation would be staying. They were expected to arrive that evening.

I again walked to the cathedral. Kathy Smalley and the CBC crew were there as well as an American Broadcasting Corporation television crew along with William Blakemore, bureau chief of ABC news in Rome. A CBC reporter was conducting a TV interview on the cathedral steps with some senior Italian prelate.

He turned out to be Monsignor Cottino, who was serving as the archbishop of Turin's press spokesman. He was reading a statement from the archbishop, Cardinal Ballestrero, to the effect that no carbon dating of the shroud would be permitted because the method was not accurate enough (he claimed that accuracies of plus or minus 200 years were the best that could be obtained) and because no one had made a formal request to do so.

Smalley got the brilliant idea of having me meet this rather dour faced man and she more or less dragged me up the steps. She introduced me as the American professor who had invented a new way of carbon dating the shroud using a miniscule sample and who was actually offering to do so. Cottino paled visibly. I thought he was going to raise his cross as a means of warding off my evil influence or draw a pistol and shoot me with a silver bullet but all he did, as the CBC camera continued to roll, was to turn on his heel and walk rapidly away. It is all deathlessly recorded on the *Man Alive* programme that aired a few weeks later. I remarked to Kathy that it was a hell of a way to win over the hearts and minds of the Turin authorities to the virtues of AMS. I had certainly started things off on the wrong foot, albeit with a lot of help from the CBC.

Kathy later told me that she had cleared the whole thing up. It was to be just the start of problems I was to have with the Turin ecclesiastical authorities and their science advisor, Professor Luigi Gonella of the Turin Polytechnic.

The truth of the matter was that I did not care whether the Turin Shroud was carbon dated or not. I merely wanted to ensure that the Turin authorities knew that it could be done in an accurate and credible way using an amount of cloth that would in no way affect the shroud's appearance. My further concern was that if it ever were done it should be done by the people who knew what they were doing and were dispassionate and not under the control of true believers.

Smalley and the CBC crew had access to a side door into the cathedral and invited me to go in with them. We could pass along the wall where the shroud was hung, well above our heads, and look up at the people passing along the wooden walkway that had been installed for viewing the shroud. As I stood looking up at the people gazing at it from the elevated walkway, I could see their rapt expressions. I suppose most of them were Italian and their emotions showed very clearly on their faces. Many were in tears.

Smalley suggested I join the line so I could be videotaped passing along the walkway looking up at the shroud. I must say I also found it a very moving sight. It is a truly remarkable object with great artistic and religious beauty. The double image on the cloth—the front and back imprint of a crucified body—is indeed very faint and the intricate details that those who have studied it claim to see are not all apparent even at the distance from which one could view it on this wooden walkway. The triangular patches along the sides but avoiding the image, applied by the Poor Clare nuns in 1534 two years after a fire burned holes through the shroud, are the most visible features. But one could also clearly see the crossed hands showing only four fingers. There were wounds at the wrists and also what appeared to be a wound in the right side, as well as ones on the head and all over the back.

I left the cathedral strengthened in my belief that the Turin Shroud was, at the very least, a religious artifact of greater significance than any other in the world that I had ever seen or heard of.

The deep emotional reaction of the people viewing it, that I had just seen at first hand, made me wonder more than ever whether there was a proper role for science to play in establishing its age and the nature of its image. Carbon dating could, indeed, establish with some degree of accuracy the year the flax from which its linen was woven was harvested. It could never prove it was Christ's shroud even if it did date to the first century. If it was substantially younger, that would prove it was not and I worried about how that might affect the people I had seen with tears in their eyes.

I walked around the centre of Turin, impressed by its beauty despite its being an industrial city. One of the largest Fiat factories is located on its outskirts. The CBC people had told me there would be a press conference held at 5 pm and that, by all means, I should attend it. Quite a crowd of reporters had gathered and they were given information on the two day congress on the shroud that was to start the next day.

I met Father Peter Rinaldi for the first time and it was immediately clear to me that he was a man of enormous charm. He was in his late sixties at the time, a courtly man, slight of build with white hair and a handsome mobile face. He first experienced the fascination of the Shroud of Turin as an altar boy in the cathedral adjoining the chapel in which it was stored. His English was excellent which was not surprising since he was a graduate of Fordham University and for many years was a parish priest in Port Chester, New York.

I asked him whether it would be appropriate for me to make a statement about our new carbon dating method and how it could be applied to the shroud. He agreed I should. He translated my comments into Italian. Immediately, some church official read the archbishop's statement that no carbon dating would be permitted. It was clear that they had an orchestrated campaign to deal with the carbon dating question. I began to feel like a pariah. At this press conference I also met Michael Thomas, a reporter for *Rolling Stone* magazine who, remarkably enough, was in Turin to cover the congress.

I walked back to my hotel and had dinner in a nearby restaurant. The British delegation arrived at the hotel about 9 pm. They included, besides David Sox, Anglican Bishop John Robinson, author of the controversial book *Honest to God* and a distinguished member of the British clergy; a noted shroud scholar, Ian Wilson, the British author who had written the best book to date on the shroud, simply titled *The Shroud of Turin*, and who had been converted to Catholicism as a result of his studies of the shroud; and a man named Peter Jennings, a British freelance journalist. Sox warned me that I should exercise care in any interviews I had with Jennings.

The *II Congresso Internazionale di Sindonologia* (*Second International Congress on the Shroud*) commenced the next day. It was held in a place called the Banker's Institute. Papers were presented in the following categories: History and Art, Medicine, Science and Technology and Exegesis and Theology. Some papers were accepted as communications and were published in the congress proceedings but were not presented orally. That was the category to which our paper 'A Method for Dating the Shroud of Turin' and a paper by McCrone 'A Current Look at Carbon Dating' were consigned. There were also several other papers mostly of a mildly lunatic variety such as one which suggested that God employed laser light or a thermo-nuclear energy source at the time of the resurrection. If so the author suggested one should look for a radioisotope silicon-30. The fact that this isotope has a half life of 2.6 hours and the author of the paper clearly believed the shroud was 2000 years old seemed not to matter.

Of the papers presented orally on the first day of the congress, probably the most interesting ones were by Ian Wilson, 'The History of the Turin Shroud'; Robert Bucklin, a pathologist and medical examiner from a Los Angeles hospital, 'A Pathologist looks

October 1978, Ken Stevenson and Eric Jumper (US Air Force and STURP) with Father Peter Rinaldi and Roy Bonisteel (CBC).

Max Frei (Swiss forensic expert) and Eric Jumper (US Air Force and STURP).

at the Shroud of Turin'; John Jackson and Eric Jumper, of the US Air Force Academy, 'Space Science and the Holy Shroud'; Max Frei, the Swiss forensic expert, on his studies of pollen taken from the shroud; and A Brandone and P A Borroni, of the University of Pavia, on the possibility of using neutron activation analysis of the

October 1978 photograph of Cardinal Anastasio Ballestrero, Professor Luigi Gonella (a nuclear physicist from Turin Polytechnic and the Cardinal's science advisor) and Baima Ballone (a Turin forensic scientist). (Courtesy of Barrie Schwortz.)

1978 Turin crowd scene in the cathedral square.

shroud cloth. The latter is certainly a sensitive way of detecting certain chemical elements at low concentrations but such information could contribute little to the shroud's provenance or to the origin of its image. Cardinal Ballestrero put in a brief appearance during the first morning, welcoming the delegates to the congress.

Max Frei showed countless electron microscope photographs of pollen samples he had removed from the shroud in 1973 by pressing sticky tapes to its surface. It was during this scientific foray on the shroud following its showing on Italian television that the cloth samples were removed for examination by the Belgian textile expert Gilbert Raes.

As a result of the pollen studies Frei made, he was reported as stating that the shroud certainly dated from the time of Christ. I was amazed when I heard of this claim. How could one determine the shroud's age from the pollen samples it contained? In his talk at this 1978 congress all he claimed was that the pollen, in many cases, came from plants indigenous to the Near East. He concluded there was a greater than 90% probability that the peregrinations of the shroud included both the Jerusalem and Constantinople areas as well as, of course, France and Italy. Even this much more modest conclusion was unconvincing to me. I have been told that expert palynologists (which Frei was not) claim he actually misidentified some of the pollen he studied.

Eric Jumper's talk mostly covered measurements he and John Jackson had made on a full-size photograph of the shroud, using a device called a VP-8 Image Analyzer System which they had borrowed from NASA. That instrument is normally used to convert shades of light intensity on moon photographs to vertical relief. They used the computer output of this device to produce a three-dimensional full scale cardboard cutout of the frontal shroud image. Jumper proudly displayed this to a marvelling audience. He concluded that the image was not produced by vapours from or direct contact with the body covered by the shroud. He and Jackson became the heroes of this first day of the congress. Personally, I did not find this evidence that the shroud image had a three-dimensional property very compelling either.

Jumper's talk did result in one rather amusing incident. There was a question in Italian from a member of the clergy in the audience that, when translated into English, was "Did he have a belly button?" I failed to appreciate the religious significance of the question until some time later, back in Rochester, as I will recount toward the end of this chapter. In any case, Jumper's answer was that the resolution of the VP-8 analysis was not sufficient to tell.

The world's press sent reporters to this meeting and I and many other delegates were constantly being interviewed. One reporter was a comely young woman named Christina

Leczinsky-Nylander, who was the producer of science and cultural programmes for the Swedish Broadcasting Corporation. She taped quite a long interview with me about the possibility of applying the new AMS carbon dating technique to the shroud. The next day I happened to mention to her that a reporter from *Rolling Stone* magazine was present. She got quite excited and asked me to introduce her to him, which I did. The next thing I knew she was interviewing him!

Jackson and Jumper, the young officers from the Air Force Academy in Boulder, Colorado, had led a large team of scientists to Turin bringing with them a considerable amount of scientific equipment. For some time they had been planning to carry out a series of measurements on the shroud. They had only very recently received permission to do so and would be swinging into action immediately following this congress when it was planned to move the shroud from its display location in the cathedral to the chapel in the Royal Palace. There the STURP team would work their 'non-destructive' space age scientific wonders on it for what turned out to be five days and nights.

Their spokesman, press liaison man and chief of security, was Ken Stevenson. Stevenson was a rather abrasive black man, a graduate of the Air Force Academy and, at that time, with International Business Machines. I had been told by a reporter from the *Los Angeles Times* during an interview he had with me that Stevenson had told him that the new radiocarbon dating method was not yet sufficiently developed to be applied to the Turin Shroud. I categorically denied the truth of this statement. I later ran into Stevenson and had a rather heated discussion with him on the subject. I suggested in no uncertain terms that he didn't know what he was talking about.

That evening back at the Venezia hotel, after dinner, I had a long conversation with the *Rolling Stone* reporter Michael Thomas in which I attempted to explain this new technique we had developed for carbon dating, how it differed from the method developed in the forties by Willard Libby and how it could establish the age of the Turin Shroud using a single thread some 20 centimetres long. Thomas later incorporated this conversation into an article which was published in the 11 January 1979 edition of the magazine. The piece was titled 'The Shroud of Turin—The First Polaroid in Palestine' and carried the subtitle 'Who is this man and why does he have no navel?'

His article is, in my opinion, the best serious but incredibly cheeky article on the shroud that has ever been written. His reaction to the STURP team is hilarious. He described them as "mad Faustian defenders of the faith dreaming up hardware that make Hiroshima look like a love bite" whose jobs in Los Alamos, New Mexico, involved "working on hunter-killer satellites and directed energy weapons like satellite-launched high-energy lasers—and particle beams that convert a couple of kilotons of underground blast into a 100 billion billion angry protons beaming through the atmosphere that will obliterate all known matter and leave no trace in the time it takes some idiot to press a green button." They were in Turin "with their glossy little wives and eerie kids" and "The reason they're here is to raise Jesus Christ from the dead."

After these devastating comments, I expected to read similarly trenchant remarks about me but he was much kinder—perhaps because we had shared a fair number of beers during our conversation. I had given Thomas what I thought was a lay description of our AMS carbon dating technique. He obviously used it because he gave a good account of the method, including its advantages over Berkeley's unsuccessful attempts to use a cyclotron. In his article, however, Thomas described my account as "a brilliant description—which I just threw in the fire." He described my reaction to the miracle of the resurrection and the possibility it produced the image on the shroud as follows: "Harry Gove rolls his eyes. He doesn't buy it. He doesn't believe in miracles. As a physicist he agrees he can't understand any other way the image on the shroud was formed except by some kind of thermal radiation. But Harry Gove's not 'born again'. He's a once-a-year Episcopalian. He doesn't believe the shroud came from the corpse of Jesus Christ. He supposes it happened some other way. And he pauses. What other way?" A footnote at the end of the first page of his *Rolling Stone* article reads 'Michael Thomas is our religion correspondent'! I suspect few people know that *Rolling Stone* has reporters in that category.

About 10:30 that evening, after my interview with Thomas, I went to the cathedral with the other congress delegates where we were given a special showing of the shroud. This turned out to be the last time I was to see it and I marvelled again at how beautiful and moving a sight it was. The possibility still remained that it was indeed 2000 years old and therefore could have been the burial cloth of Christ and the imprint on the cloth the actual image of his body. Only carbon dating could answer the question

ROLLING STONE, DECEMBER 28, 1978–JANUARY 11, 1979

THE FIRST POLAROID IN PALESTINE

✐✐✐✐✐ The ✐✐✐✐✐

SHROUD of TURIN

Who is this man and why does he have no navel?

BY MICHAEL THOMAS

THEY'RE not much to look at. Lounging round the swank Hotel Sitea in Turin, Italy, this fall in their burnt-umber double-knits and earth shoes, they look about as dangerous as a game of golf. No dandruff here. No unruly old Jewish scientists with wire-wool sprouting out of their bald spots like shot fuses and no egg on their ties, no ramrod Prussians with frosted eyes and dueling scars, no Strangeloves. One guy looks like Jackie Gleason. Otherwise they're about as charismatic as cardboard. But make no mistake. These boys are going to bury you.

This is what's known in nuclear circles as the scientific community, and this team in their Johnny Miller strides with their glossy little wives and eerie kids are the frontline. They have rays and beams and megatonnage and plagues of leprous infection aimed at every enemy of the U.S. (and that means the whole world and the galaxies beyond, except for New Mexico and Colorado and maybe California) that make 'Star Wars' look like bows and arrows.

There are a couple of hundred (overtrained) blue-eyed boys in Los Alamos, New Mexico, working on hunter-killer satellites and directed energy weapons like satellite-launched high-energy lasers—and particle beams that convert a couple of kilotons of underground blast into 100 billion billion angry protons beaming through the atmosphere that will obliterate all known matter and leave no trace in the time it takes for some idiot to press a green button. There are physicists and chemists and engineers and mad Faustian defenders of the faith dreaming up hardware that make Hiroshima look like a love bite. There are sheer genii out at Los Alamos with security clearances so high they dare not even talk to themselves.

I could tell you their names, but they don't want to be disturbed. Every time I just idly happened to bump into a thermal chemist in the bar at the Sitea and just nonchalantly asked him something innocent about say the aloes and myrrh used to marinate Jesus in, one of his mates would materialize from behind an aspidistra and whisk him off before he could loosen his string tie and clear his throat. But they're all here—thirty or forty of the best brains in the Jet Propulsion Laboratory and Sandia Laboratories and the Air Force Weapons Laboratory—hanging round the Sitea with half a million dollars' worth of microdensitometers and x-ray fluorescence streptoscopes and probes and scans and lasers and, God knows, probably a couple of overtrained German shepherds in crates in Milan.

The reason they're here is to raise Jesus Christ from the dead.

MICHAEL THOMAS *is our religion correspondent*

Thomas' article in the December 1978 issue of Rolling Stone. (Courtesy of Michael Thomas and Rolling Stone.)

and if it ever were carried out, that would be the result I would have preferred. Thomas' description of my religious beliefs was correct. However, Christ was a real and tremendously influential person and it would be much more interesting if it turned out the shroud might have been his burial cloth.

One of the best papers delivered on the second day was by John Robinson titled 'The Shroud and the New Testament'. Admittedly the papers that impressed me most were the ones given in English since I have no familiarity with Italian, but there was simultaneous translation into English of the other papers and the ratio of factual information to religious interpretation of most of their content was depressingly low. Most of the speakers fell into the category of 'true believers' that the shroud was the burial cloth of Christ—including, I fear, Bishop Robinson. His was a very scholarly presentation, however, and I learned a lot from it in terms of the remarkable match between details of the image on the shroud and the scriptural description of Christ's crucifixion.

At the end of his talk, Bishop Robinson made a strong plea for having the shroud radiocarbon dated by the new method that required such a small amount of material. Despite my requests to the congress organizers, I had not been given an opportunity to describe even briefly how this new carbon dating could be accomplished. This was the chance I was looking for—and none too soon—the congress was about to end. I joined the line of people waiting to comment on Robinson's talk. Ian Wilson offered to make some remarks and then to introduce me—an offer I was pleased to accept—after all, Wilson was one of the stars of the congress. Before he could do so however, some Turin cleric, who had recognized me, again read the archbishop's statement that carbon dating would not be permitted! It struck me that this was becoming analogous to throwing holy water on the devil!

Wilson then made his introduction and as I moved to the podium I could see considerable press and TV camera activity. When I reached the podium it was as if I were some sort of Hollywood celebrity. The TV camera lights came on, flashbulbs popped and the TV cameras rolled, including those of the CBC's indefatigable Kathy Smalley. When I later saw the CBC *Man Alive* programme I noted that my brief moment of glory was faithfully recorded for the pleasure of the viewing public—and, I must confess, mine.

In my remarks, I briefly described how we would go about dating the shroud using this new method and stressed the miniscule sample size required. I said it was perfectly agreeable to me to wait a year or so until another laboratory was also in a position to collaborate, but I thought it important to let the congress delegates and the Turin authorities know that we could do it now if Turin wished to have it done. During the break that followed, I was besieged with questions by the press. Their attention outrivalled what they had bestowed on Eric Jumper after his show and tell involving the cardboard cutout of the shroud image. I thought to myself that finally these spit and polish young Air Force officers with their vaunted space age technology were being outshone—and by a professor of physics old enough to be their father, who just happened to have superior technology at his command.

Lunch, this second and last day of the conference, was held in some scenic picnic area on the outskirts of Turin where we were taken by bus. It was the official congress sponsored lunch and a typical Italian wine drenched affair. I purposely sat beside the STURP public relations man Ken Stevenson to try to establish a more congenial relationship between us. He is very articulate, a religious zealot and, like the vast majority of his fellow STURP members, a 'true believer' that the shroud was the burial cloth of Christ. Like Jackson and Jumper, he had an exaggerated view of the importance of this mission, the overwhelming need for security and secrecy, and of his own importance in STURP.

He told me the fact that the safe arrival in Turin of some 40 members of the US STURP team including wives and children, with all their boxes of fancy measuring equipment, was 'a miracle from God'. I smiled at this because this statement of his had been gleefully reported to me the day before by the Swedish Broadcasting Corporation reporter. According to Stevenson, not one penny of the money spent on their enterprise had come from government funding. It was all from private sources— except, I assume, the loan of such devices as NASA's VP-8 Image Analyzer system. The ameliorative properties of good Italian wine contributed to achieving an amicable relationship between us—at least for the nonce.

In the afternoon session, David Sox conveyed to the congress greetings from the British Society for the Turin Shroud. He stressed the importance that society attached to having the shroud carbon dated. A few more speeches followed of a religious nature. There had been far too many of these already, to no purpose and far too emotional, and then the meeting was ended.

That evening, Bishop Robinson and I were invited to dinner by Kenneth Weaver, senior assistant editor for the *National Geographic*. They had been given the exclusive rights to cover the tests STURP were to carry out in the succeeding days and to be the first to publish the findings when STURP made them available. They did so in their issue of June 1980 in a really splendid article. At the end of the article Weaver discussed the issue of carbon dating and quoted me as saying that with one square centimetre (the size of the tip of one's little finger) I can provide an age conservatively accurate within 150 years. He concluded that Turin's archbishop will eventually give permission for carbon dating. 'Eventually' turned out to be ten years. The sample size was correct, the accuracy quoted was conservative—it turned out to be within 65 years—but it would not be me who did it.

I had a brief chat with David Sox later that evening. He said there was a rumour that McCrone would get samples of the shroud. He and his wife Lucy (who had been at our April meeting in Rochester) had attended the congress. He was saying that Berkeley would not be ready for six months or so. Sox thought that samples would be taken during the STURP tests and would be made available for carbon dating when two laboratories were ready to take on the job. It seemed clear to me, after the repeated pronouncements that the archbishop was not ready to allow carbon dating, there would be no samples taken this time around.

I was up at 6 am the next morning to drive to the Hotel Sitea where a limousine would take me to Turin airport for my flight to London and then home. I had just finished my shower when the phone rang. It was Christina Leczinsky-Nylander of the Swedish Broadcasting Corporation wanting to know whether there were any new developments on the carbon dating front. I told her there were not but I thought there was a good probability that we would ultimately get a sample.

The Sitea is a fancy and very expensive hotel where the better heeled congress delegates stayed, including most of the STURP team along with their wives and children. Bishop Robinson had accompanied me and we went in to the hotel restaurant to have a cup of coffee. The American medical doctor, F Zugibe, an expert on the deleterious effects of crucifixion, whom I had previously met, was having breakfast there with his wife. I introduced Robinson to him and the two had a spirited conversation on the question of rigor mortis, when it would set in and when it would be relieved in the case of a crucified man. I thought it was a remarkable subject to discuss at breakfast.

My Pan Am flight out of London's Heathrow was delayed so

I missed my connection at Kennedy. Pan Am put me up at the International Hotel there. I phoned my wife in Rochester to say I would not arrive until early afternoon the next day. She said Ken Purser, his wife and his youngest child were staying at our house and that the Rochester TV reporters had been at the Rochester airport to interview me. She said Ken would delay his return to Boston so he would have a chance to learn about my Turin experiences. It was Ken's company that had financed my trip to Turin.

The next day, I arrived in Rochester around 2 pm and was met by reporters from the three local TV stations and the two local newspapers. I was beginning to feel like a real celebrity—something that happens very infrequently to professors of physics. They wanted to know what had transpired in Turin and whether I had samples from the shroud to carbon date. I put on an optimistic front and told them—and I really did believe it—that samples had not yet been made available for carbon dating but I thought it would not be long before they were. I drove my wife home and then went to my lab to meet with Purser. While there I did another interview by phone with Jack Jones, a reporter from the local morning paper I had talked to on several previous occasions.

It was during a talk I gave a year later in the main library of the University of Rochester to a group of undergraduates that I commented on the question 'Did he have a belly button?' which had been raised on the first day of the Turin congress after Eric Jumper had shown his three-dimensional cardboard cutout of the shroud image. I said that I wondered what its significance was. After the talk, a young woman in the audience went to a nearby bookshelf which contained the *Catholic Encyclopaedia*. It discussed the question of Jesus' virgin birth and stated that not only was the conception immaculate but Mary sustained no evidence that a birth had occurred and she remained a virgin to the end. This seemed to preclude an umbilical cord or placenta and thus there would be no navel. In discussing the matter with one of the leading Catholic prelates in Rochester, he pointed out that the Bible states that Christ was born of man with presumably all of man's accoutrements which surely would include a navel.

It occurred to me that the question of whether Christ possessed a belly button might make an interesting research project. But where would one go to obtain the requisite funding to carry out such a study? I doubted if there were an appropriate division of the National Science Foundation that would be interested. At that time the University of Rochester provided management services for a Washington based federal agency. For these services the university received a management fee which was used to support research in areas not normally eligible for federal grants, especially those in the humanities. It struck me as an ideal source of funds for research on the question of whether Christ had a navel. The organization in question was called 'The Office for Naval Analyses'. After all, it required the change of only one letter. I wrote to my dean, who was normally not lacking in humour, requesting support for the project but never received an answer. I had occasion some time later to meet the man in charge of this office and told him about my request. He was not amused.

From 13–15 November 1978 I attended the Council for the Advancement of Science Writing, Inc.'s meeting on New Horizons in Science which was held in Gatlinburg, Tennessee. I had been invited to give a talk on our AMS work and the possible application to carbon dating the shroud. I think the invitation had been suggested by Ben Patruski, who was writing the AMS article for the NSF's magazine *Mosaic*. The main purpose of the meeting, attended by science writers throughout the country, was to have talks by people involved in all the important science stories that had made the news that year. I announced at the end of my talk that the Raes samples would be made available as soon as two laboratories were ready to take on the dating job. This caused great excitement among the science writers present, many of whom wrote for various newspapers and, as a consequence, the talk got considerable press coverage.

I got back to Rochester around 9 pm on the 15th of November. The mayor of Rochester was on the same plane. As I walked into the airport lobby from the gate, I was greeted by a reporter and cameraman from the Rochester television station Channel 10. They started filming me as Mayor Ryan walked by. He looked a bit quizzical—clearly he was much more used to this sort of greeting than I was and he must have wondered who I was to be taking the spotlight from him. The reporter wanted to know when we would get samples from the shroud. I suppose I made the usual optimistic reply that it would be quite soon. It was all very heady stuff. I began to think that maybe the best thing that could happen to me and my laboratory would be to continue getting all this free publicity without ever having to actually date the shroud.

Chapter 3

Interactions with STURP

In late November 1978 I attended the Materials Research Society's annual meeting in Boston where I gave an invited paper on our AMS work. There I met Ed Sayre of the Boston Museum of Fine Arts. Sayre also had a connection with a group in the Chemistry Department of Brookhaven National Laboratory with whom he was collaborating on the development of small proportional counters for carbon dating by the Libby decay counting method. These counters were specifically designed to deal with small samples. We discussed a possible collaboration between Rochester and Brookhaven to date the shroud.

When I returned to Rochester I called Ray Stoenner at Brookhaven. He and Garman Harbottle were the people there involved in the small-counter work. Stoenner expressed great interest in being involved in an effort to date the shroud—he said he could start tomorrow! With 10 milligrams of carbon from a 2000 year old sample he could achieve an accuracy of ±115 years. It would take a couple of months of continuous counting to do so, however. It would also take an equal amount of time to measure a background and a known standard but these additional two measurements might be made simultaneously with duplicate counters.

Soon after talking to Stoenner, I spoke to Sox and mentioned such a collaboration. It made a lot more sense to me than a collaboration with Berkeley. Our interactions with Berkeley, in any case, made such a collaboration problematical at best. Sox said that he had just received a letter from John Jackson, the STURP leader, who told him that Ray Rogers, a Los Alamos chemist and a member of the STURP team, had suggested to Jackson that the Brookhaven group be involved in the carbon dating. I thought this was quite a coincidence. However, it turned out that Rogers and the other STURP member at Los Alamos, Robert Dinegar, knew Harbottle. Sox suggested we send a joint proposal to Cardinal Ballestrero with a copy to Peter Rinaldi. I said I thought I should contact Jackson first. Sox said that he had heard from Eric Jumper (Jackson's airforce colleague and a co-leader of STURP) that no sample had been removed from the shroud for carbon dating after the 1978 congress and the subsequent STURP measurements, but I should double check with Jackson.

In early December I phoned Jackson at his home in Boulder, Colorado. It was also listed as the STURP headquarters. I must have spent an hour and a half talking to him. He told me the 'American team' that included McCrone had signed an agreement to finish the analysis and the written report of all the tests they carried out last month by October of 1980, i.e. in about two years, and not to leak anything before then. He said he hoped the results would be ready earlier. They would then send the information to Cardinal Ballestrero for release. Jackson said that, of course, he would like any carbon dating results that might be available by then to be included. He recognized, however, that the groups doing such dating could proceed independently if they chose.

I mentioned the suggestion that the carbon-14 measurement be a Rochester–Brookhaven collaboration and that we should prepare a joint proposal. He was enthusiastic about this idea—particularly since it would combine the traditional Libby decay counting method with the new direct detection AMS method. Such a combination might appeal to Turin. He said that Ray Rogers might have already proposed to Turin, through Peter Rinaldi, that Brookhaven be involved. He suggested that our joint proposal should be sent to Ballestrero via Rinaldi.

He said the STURP team worked day and night for five days on their tests of the shroud following the congress. When I asked him if any cloth samples had been removed he said only a one centimetre long thread had been taken in the area of a 'blood' stain. He said a man named Giovanni Riggi had vacuumed the shroud to collect superficial dust and pollen samples and also that some really excellent photographs had been taken. He said it would really be obvious if a sizeable weft thread were removed and so he doubted if there would be any chance of getting such a thread, especially across the image, for carbon dating. He pointed out that the Poor Clare nuns, in 1534, had applied patches on both the front and back of the shroud over the burned areas. Clearly there was lots of material under these patches, the removal of which would not be visible. We ended a lengthy, informative and friendly conversation.

On 26 December I received a postcard from David Sox. He had talked to the former king of Italy, Umberto II, who was visiting London, and said the king was enthusiastic about the possibility that the shroud would be carbon dated. Well before the dating finally took place, the king would be dead and the shroud willed to the Vatican. At least we knew he would have approved the dating if not its result.

A January issue of *Science News* carried an article quoting Richard Muller (the leader of the AMS group at the Berkeley cyclotron) as saying that the AMS carbon dating method would not be ready to tackle the shroud for another year. He meant he would not be ready until then, but he was audacious enough to imply that all of us in the AMS field were similarly incapacitated. McCrone was quoted as saying "Rochester was ready yesterday".

On 8 January 1979 John Jackson contacted me. He said that there would be a meeting of STURP in Santa Barbara, California on 24 and 25 March 1979, to make a preliminary analysis of the data that was collected in October 1978. He said he would be pleased to have me attend.

I talked to David Sox, who was now in California, and told him that the CBC had learned that no shroud thread would be destroyed. He said this was undoubtedly the work of Monsignor Cottino, the priest to whom Kathy Smalley of the CBC had tried to introduce me on the cathedral steps in Turin in October 1978. Sox recommended that our dating proposal should be sent instead to Cardinal Ballestrero via Don Piero Coero-Borga and gave me his Turin address. He said Rinaldi was now in the USA at Corpus Christi Church in Port Chester, New York.

I phoned Ted Litherland in Toronto regarding the offer we were preparing for Turin to date the shroud. He suggested we make it clear that dating the shroud would not take precedence over other measurements in which we were involved. It is interesting that he was never as enthused as I about becoming involved in the shroud. He may have felt, like Robert Hedges, that religious artifacts were not worth wasting one's time on. I agree that is generally true but the Turin Shroud is a rather special case. He said he expected to hear within the next ten days about his proposal to create an AMS laboratory at Toronto based on one of Purser's Tandetrons.

On 16 February 1979 (a typographical error gave the year as

1978) I mailed a letter to His Eminence the Archbishop of Turin, Monsignor Anastasio A Ballestrero, c/o Don Piero Coero-Borga on behalf of the Rochester, Toronto, General Ionex, US Geological Survey group and the Brookhaven National Laboratory group, offering to date the Turin Shroud (in this letter I was careful to capitalize the word shroud—even when it was not designated as Turinese). We would use two methods, AMS and conventional decay counting using milligram samples of cloth. Enclosed with the letter were descriptions by each group of the procedures they would follow in carrying out the task.

Copies were also sent to Rinaldi and Sox. It was at the latter's suggestion that it be sent to the cardinal via Coero-Borga but the latter turned out to have been an unfortunate choice. We later learned Coero-Borga chose not to deliver it to the cardinal/archbishop. (Ballestrero was the archbishop of the Archdiocese of Turin—an administrative post. He had also been elevated to the position of cardinal. He was sometimes referred to simply as 'the bishop'.) Meyer Rubin was a co-author of our half of the proposal even though he remarked wryly that if the shroud's age turned out to be substantially less than 2000 years, Christians would say "Those Jews have done it to us again!". The other authors on our proposal were D Elmore, R P Beukens, and, of course, Litherland and Purser and me. The Brookhaven proposal was co-authored by G Harbottle, R W Stoenner and E V Sayre.

The letter to Ballestrero read:

"Your Eminence: Enclosed you will find descriptions of two independent methods that could be employed to establish the age of the Shroud of Turin. The one (from Rochester) describes the new tandem electrostatic accelerator technique that was first employed at the University of Rochester in the direct detection of radiocarbon from contemporary and ancient carbon samples in May 1977. It has been perfected since then to the point where it can date milligram samples of carbon in the age range 1000 to 2000 years to an accuracy of better than plus or minus 160 years. The other (from Brookhaven) describes a method developed at the Brookhaven National Laboratory which employs specially designed and shielded proportional counters and which can date samples containing as little as 10 milligrams of carbon with an accuracy similar to that stated above. It is a refinement of the method developed by Libby.

"The two laboratories, one (Rochester) employing the new

technique and the other (Brookhaven) employing the conventional method, would operate completely independently from sample preparation to final determination of the age of the sample. The results from one laboratory would not be made available to the other at least until both laboratories had completed their measurements and determined a date.

"This letter constitutes our formal offer to date the Shroud of Turin, if you desire to have it done. We can discuss additional details of the procedures to be followed if you desire to proceed further. We could begin immediately to prepare to make the measurements described in the two documents. At both laboratories, however, this project would have to be carried out in a way that was compatible with other commitments.

"We are taking the liberty of sending copies of this letter to the Reverend Peter Rinaldi, S.D.B. and the Reverend David Sox."

I signed it on behalf of both groups. We never received either an acknowledgement or any other response to this offer because it was never delivered to the cardinal.

In the Rochester consortium proposal we considered the amount of cloth required. We discussed the lengths of both warp and weft threads that would be required, but lacking a good figure for their respective lineal densities or the conversion factor between the weight of carbon contained in a given weight of cloth, our estimates of the amount of material were crude at best. We did state, however, that the equivalent amount of carbon we would need was between 5 and 10 milligrams. We also noted that the Raes samples might be used or uncharred material taken from underneath the patches. Luigi Gonella, the archbishop's science advisor, later accused us of favouring the Raes samples. Even though, at the Santa Barbara meeting described below, he suggested our use of them, he later claimed they were suspect, and this was the reason our offer was not taken seriously. That was patently a misstatement of fact. We learned later the Raes samples had been so casually treated by Turin ecclesiastical authorities that questions might have been raised concerning their authenticity. On the other hand, the shroud weave is so unmistakable that identification would have been certain.

In February I wrote to Sox and Jackson saying that I would attend the shroud data analysis meeting in Santa Barbara on 24 and 25 March 1979. I thought it would be an opportunity to discuss the shroud carbon dating offer made jointly by

Brookhaven and the Rochester group in less hectic circumstances than prevailed in Turin.

On 22 March I left for Los Angeles to attend the meeting. I stayed overnight with Sox in a house he was renting in Costa Mesa. We were to drive together the next day to Santa Barbara. He had a teaching job at the Harbor Day School somewhere in the vicinity and would be there most of the day. He showed me three letters he had recently received which he said I could read but not photocopy. Since he would be back about 3 pm, I asked him whether I could use that time to copy the relevant sections into my diary and he said I could.

The first was a letter Sox had received from Father Peter Rinaldi sometime within the last month. In it he said that he would be at the Santa Barbara meeting and had heard I would also be there. He said that he had heard from Turin that I had precipitated matters there in a way that displeased Monsignor Cottino. Cottino felt pressed on the question of carbon dating the shroud and had overreacted by suggesting it would be better to wait ten years or so. He wanted STURP to decide on the carbon dating question.

Rinaldi then wrote that it had been reported to Turin that in a lecture or press conference I had said: "Think of all the friends we agnostics will have if we can establish that the shroud is only 600 years old...". He described this as a very unprofessional remark—well calculated to turn Cottino off and lots of other people as well.

I assured David upon his return from school that I made no such statement. I have always been very careful, and was particularly so in Turin in 1978, to indicate that I felt the odds were against the shroud being Christ's burial cloth but that I really did hope it would turn out to be 2000 years old. Any such statement as the above would be completely contrary to my real feelings. I suppose my ill concealed disdain for those scientists who were 'true believers' did not gain me many brownie points amongst the majority of the people involved in the shroud and that one of them had decided to spread this falsehood.

The second document, a submission to Turin by Walter McCrone titled 'A Current Look at Carbon Dating', was written in October 1978. It gave McCrone's opinion of the current status of the field. It listed Brookhaven, the University of California at San Diego, and the National Bureau of Standards as small-counter laboratories (he was wrong about the latter two) that require only 20 milligrams of linen (about 1 square centimetre) and the accelerator groups included Rochester, Simon

Fraser University in British Columbia, Canada, the University of Oxford in England, Atomic Energy of Canada in Chalk River, the University of Pennsylvania and Berkeley in the USA. He said it seemed reasonable that by December 1978 there would be two laboratories (Rochester and Berkeley) that would be able to date a few milligrams of linen with an accuracy of better than 10%. He thought that the accuracy might be better at Berkeley but that Rochester would be able to date somewhat smaller samples. It is worth noting that McCrone was so unschooled in carbon dating he did not realize 10% accuracy meant 800 years!

He said that the two samples removed for study by Professor Raes weighed about 50 and 100 milligrams respectively. He described how he would reduce the Raes samples to carbon dioxide gas—seal it in a glass ampoule for use with a cyclotron, or reduce the gas to elemental carbon and prepare a pressed pellet for the tandem accelerator.

He suggested a protocol—in which he, of course, would play the key role. The measurements would involve control samples as well as a shroud sample and would be carried out blind. The person (McCrone, naturally) authenticating the shroud samples as the same ones studied by Professor Raes should maintain the further chain of evidence so that he (and only he) knows for certain the identity of the sample analysed by the dating laboratories. The results from the laboratories would be delivered, still sealed, to the proper authorities with the key to the identities of the samples. In this way only the person opening the sealed envelopes and comparing the data therein would be able to determine the age of the shroud samples.

He then 'beseeched' Turin to release the Raes samples to him and ended by pledging that the dating analysis would be done blind and that only the final authorities to whom the dating analysis and the key to the samples are transmitted would know the true shroud date.

My reaction to McCrone's 'beseechments' is predictable. He knew little, if anything, about carbon dating. In my view there was no reason for him to play any role in dating the shroud.

The third document was a letter dated 14 March 1979 from Rich Muller (our Berkeley AMS competitor). Muller stated he would be in business sometime in the summer. He said he did not think any laboratory could claim an ability to date the shroud until they had successfully completed a series of blind measurements on samples no larger than they will get from the shroud. As of then, no laboratory had done this. He assumed this standard would apply, rather than that of simply taking a laboratory's word that they were ready. He said once some improvements in his cyclotron ion source were made tests on blind samples would be the first order of business. He enclosed a preprint of an article describing a correct blind date he had been lucky to obtain. He did not mention a second blind date he had attempted and which was badly wrong. I had managed to persuade him to add it as a footnote to an article that appeared in another journal.

Each of these three documents contained statements that I found profoundly objectionable. One contained the false charge that I had publicly stated that I, and my fellow agnostics, would rejoice if the shroud were younger than 2000 years. The other revealed McCrone's attempts to become the manager of the dating enterprise—a role I believed he was totally ill equipped to play. The third was Muller's statement that no AMS laboratory should consider itself ready to date the shroud unless it had, like Berkeley, successfully carried out blind measurements. In point of fact, his lack of success in blind dating demonstrated Berkeley's unsuitability for the task.

Later that afternoon, Sox and I drove to the Brook's Institute in Santa Barbara to attend the first data analysis workshop on the Shroud of Turin. It was to begin the next day.

Before I was allowed to attend this workshop, I had been requested by John Jackson, at that time president of STURP, to sign a non-disclosure agreement. It was designed to ensure that I did not reveal to a breathlessly awaiting world the astounding revelations that would be forthcoming at the workshop. I did so after making some minor changes in wording that ensured the agreement would only apply to results obtained by the STURP team (of which I was not a member) and not to any possible carbon dating results. To seal this silly agreement I enclosed a check for $1.00 made out to Jackson.

The first session, on Saturday 24 March 1979, chaired by Eric Jumper, was a presentation of photographic data. There was an opening statement by Jackson who noted that the two representatives of *National Geographic* who had been present at the 1978 tests in Turin, Ken Weaver and his technical assistant Mickey Sweeney, were also here at this meeting.

Vern Miller, a member of the Brook's Institute and who had been part of STURP's photographic team, discussed the photos he had taken using various wavelength filters. Certainly one of the most successful of the activities carried out by STURP in 1978 were the photographs taken of the shroud. They were on display in the halls and corridors of the institute. An excellent collection of them had also been donated to Turin and is now displayed at the entrance to the chapel where the shroud is stored. I saw them when I visited Turin in 1986 to attend the workshop on procedures to date the shroud.

The two sessions after lunch, chaired respectively by Jumper and Jackson, were devoted to microphotography, X-radiography, X-ray fluorescence, infrared measurements, analysis of particulate matter picked up on sticky tapes and ultraviolet and visible light spectroscopy.

Sam Pellicori showed some spectacular colour microphotographs, including a known burn mark from the 1532 fire that went all the way through the fibers, and evidence that the image itself was superficial. In the 'blood' areas there are clumps of reddish-yellow material in the crevices. The 'blood' penetrates right through the cloth. Under ultraviolet light neither the 'blood' stains nor the image fluoresced but scorches, water stains, candle wax and background did. Irish linen scorched with a hot iron and with acids also fluoresced at these wavelengths.

During this whole meeting, I had been taking notes as if what this motley mixture of scientists, priests, ministers, and peacetime warriors were reporting provided significant information regarding the real question of the authenticity of the shroud. They seemed to me to be a group of kids playing with expensive toys, hoping they would reveal some ultimate truth—a truth of which most of them were already convinced. I summarize it to show the kind of information STURP was collecting and how little it mainly signified.

The next topic was computer enhancement and analysis of data photographs. We were informed that no computer work had been done at the Jet Propulsion Laboratory at Caltech because they had been busy with some current aspect of NASA's Jupiter programme. I was heartened at the knowledge that real science took precedence over the shroud at JPL.

Jackson showed a set of biblical coins and suggested that it would be very important for JPL to work on the eyes. This was triggered by the suspicion that coins had been placed on the crucified man's eyes and if they could be identified it could establish a date. Since neither the photographs nor, as we have seen, the three-dimensional analysis provided sufficient evidence to settle the question of whether the man had a navel, it seemed unlikely that one would be able to establish the presence of coins, much less to identify them.

Luigi Gonella, the science advisor to the archbishop of Turin whom I had met for the first time in 1978 in Turin, was also present. Even at this time he was in favour of carbon dating the shroud. In a discussion I had with him during lunch, he told me that he thought our carbon dating proposal had been blocked by the man Sox had suggested we submit it to, Don Piero Coero-Borga. That subsequently turned out to be true. Don Piero did not even bother to show it to Ballestrero!

After lunch the bombshell came from Walter McCrone. Walter has always had a flair for the dramatic. I have no idea how competent a scientist he is, because his field is so different from mine—but he has great stage presence. His turn had come. He arose and announced "Anyone who is emotionally involved in the shroud had better relax their emotions". Since this included virtually everyone present, it was a statement guaranteed to command instant attention. He stated that he had detected copious amounts of iron oxide on sticky tapes pressed against portions of the image and especially on the 'blood' stains but none on the bare cloth. He claimed to be genuinely concerned at this large amount of iron oxide and stated categorically that iron oxide cannot arise from haemoglobin in blood.

He clearly implied that he believed that the image was a result of the application of iron oxide. This, of course, could explain the fact that the image was impervious to the high temperatures that had occurred in the fire of 1532, but there was an alternative explanation which Walter chose to ignore. Maybe several hundred years ago someone was concerned about the extreme faintness of the image and decided to touch it up. What better pigment to use than finely powdered iron oxide? The same reasoning could have been applied to Walter's other public triumph—the Vinland Map. It was his discovery that some of the map's white pigment was titanium oxide, a modern compound, that made him conclude it was a fake. Again it could have been the result of a more recent touch-up job.

John Heller, a New England medical doctor and an authority on blood, claimed that it should be possible to establish whether the 'blood' stains were really blood and even whether they were human blood, but at that time he had not done so. I sat beside Heller the next evening at the final workshop dinner and he privately remarked to me that "McCrone is an iconoclast—a balls crusher".

Throughout the whole of the day's discussions I kept wondering to myself why Jackson, Jumper, and another member of the STURP team were all wearing crosses around their necks. Hardly evidence of the dispassionate scientists they professed to be. So far as I knew, they were neither priests nor ministers. One, of course, should never knock piety, but its ostentatious display by these shroud scientists did nothing to recommend their scientific detachment. I suppose if I had not been on my best behaviour I would have baldly asked them the reason for this Christian ornamentation, but I refrained.

The next day was Sunday and it began at 7 am with a religious service led by Father Rinaldi but in which David Sox and John Jackson also officiated. Someone commented after the service that there was some 'Egyptian' cloth in Turin, presumably in the Egyptian museum there, that has the identical herringbone weave to that of the shroud and it dates to 137 AD. Ray Rogers said this was not so. He claimed that no linen dating back that far was of such a weave—it was all basket weave.

During the morning we visited Barrie Schwortz' photographic shop to look at his collection of 35 mm colour slides taken during STURP's five day examination of the shroud in 1978. We were allowed to order duplicates from him except for some that were deemed not to be in the public domain. I ordered half a dozen or so and regularly use them in talks. One I find very amusing is a shot of Jackson and Max Frei—the Swiss forensic expert— examining the shroud stretched out on its specially designed rack. Frei is about to apply a piece of sticky tape to the surface of the shroud to pick up pollen samples. Jackson is again wearing a rather large wooden cross around his neck.

Back at the institute, the meeting reconvened. Gonella told us the present scientific commission appointed by the archbishop of Turin would continue to serve as the arbiters of future tests. However, it would play no management role in them and would exercise no censorship—the results of any measurements can be freely published anywhere. The archbishop's permission is not required to publish any data obtained on the shroud.

It was my turn on the programme and for the first time I was able to describe to the members of STURP what the new method of carbon dating by AMS was all about. I stressed that the advantage of AMS was the sample size required—at least a thousand times smaller than that required for conventional decay counting. I showed some examples of measurements we had already made and readily admitted that, except for the baby woolly mammoth, none of these had been carried out 'blind', as our competitor Rich Muller was so vociferously and gratuitously advocating. I probably indicated fairly directly in my talk that I believed that carbon dating was the only scientific test that was worth carrying out on the shroud at the present time. Such a measurement, that could now be carried out using a negligibly small sample, would settle once and for all whether the shroud could have been Christ's burial cloth or whether it was of more recent vintage and thus either a relic or an icon (or a deliberate hoax). The nature of any further tests should be predicated on the shroud's age.

During the break following my talk, I had a long conversation with Gonella as we strolled back and forth through the grounds of the institute. He was amazed to learn that McCrone was not in the carbon dating business. He said he wanted to contact the other AMS laboratories and I encouraged him to do so. In particular, I said he should contact Berkeley directly and, just as he should not take my word that Berkeley was not ready to date the shroud, he should not take Muller's word that we were not ready.

We discussed how the sample should be delivered—at that stage he was suggesting the use of the Raes samples. Years later he would ridicule me for suggesting that these samples be used, because they were suspect—their history under Raes' control was undocumented. If that were so, it was because the Turin authorities had been so incredibly lax in keeping track of them.

His tentative suggestion was that one half of the Raes sample removed from the main body of the shroud should be brought by special courier from Turin to Rochester. We would remove what we needed and then send the remainder to Brookhaven. It seemed an eminently sensible and straightforward arrangement that would have produced a date for the Turin Shroud without all the subsequent ridiculous complications.

In early April 1979, I visited Los Alamos National Laboratory to give a colloquium on AMS and the possibility of dating the Turin Shroud. While there I met with Ray Rogers. Rogers believed, on the basis of preliminary analysis of the data STURP amassed, that the image was not the result of a pigment nor a result of any body fluid phenomenon. He felt the evidence continues to suggest a scorch because of its superficial nature. The brown colour is on the top 10 to 20 microns of the high points on the cloth.

I suggested that the image could have been formed by some artist taking a flat bronze plate and etching the image on it like a bas relief, heating the plate and then pressing it to the cloth. Rogers said that nobody had yet made that suggestion and he seemed quite taken with it. The 'blood' stains could then have been added using real blood—a dedicated artist could even have pricked his own veins. One problem with this idea that Rogers raised was that the image was on top of the blood stains. However, that would have just required the forger to apply the blood first and the hot metal plate second.

Rogers did not consider it significant that the image had the characteristics of a negative. For example, the positive of the frontal view of the face would show highlights on the nose and cheeks with darkening as one moves toward the side of the face. A negative would show the nose and cheeks dark with a lightening as one moves toward the sides of the face. This is exactly what a scorch of a face would look like. The parts closest to the cloth would be darkest.

He recounted many fascinating stories about various bizarre happenings during the five day test period. For example, late one evening some people brought a full-size replica of the shroud into the room, unrolled it, pressed it up against the shroud, hurriedly rolled it up, and rushed out before they could be apprehended. Others, from time to time, brought in various objects and pressed them against the shroud in hopes that would impart special properties to them.

In mid April back in Rochester, Kathy Smalley of the CBC phoned. She said that Monsignor Cottino had been quoted in a recent issue of the *British Broadcasting Corporation (BBC) Times* as stating that no carbon dating had been requested. That squared with the information that I had from Gonella at the Santa Barbara workshop that the Rochester–Brookhaven offer had been consigned to limbo by Coero-Borga. I passed this information on to Smalley. In the *BBC Times* article, Cottino stated it would take too long, it would require too large a sample, it was not accurate enough, and so on. Clearly the real reason was that he feared what the results might be.

Shortly after this I attended, at Brookhaven National Laboratory, a meeting of the Associated Universities, Inc. of which I was a member of the board of directors. At the dinner the first evening, I sat at a table with Ed Sayre and his wife. He said the Brookhaven small-counter carbon-14 group had run samples of mummy wrappings and things looked good. The next day I met briefly with Harbottle. He told me that they had followed Rubin's procedures for cleaning linen and the sample had disintegrated and then dissolved. One could just picture being reduced to gibbering idiots if this happened while we were cleaning the shroud sample! I suggested that Rubin and I should meet with him as soon as possible at BNL to reach agreement on the proper cleaning procedures.

In late April 1979 I was contacted by Seth Hill, who was associated with the programme *In Search of* hosted by Lenard Nemoy (aka Dr Spock, of Star Trek fame). I had heard that they were planning one of this series on the shroud and wanted to include something on the new dating technique that could be applied to establish its age. Hill said he would be coming to Rochester on 3 May to film a sequence to be used in the programme and that a camera crew from the local CBS TV station Channel 10 would do the filming. I assumed Nemoy would also be coming. That would be quite exciting.

Robert Dinegar wrote to me stating that, at the request of the Turin authorities, STURP had appointed a committee on radiocarbon dating to recommend to the archbishop what action should be taken in this area. This committee would welcome a short research proposal from Rochester on how we would propose to date the shroud. I should also indicate the expected accuracy, the quantity of material needed and how long after receiving the samples it would take to obtain a date. He said that they were hoping to have this carbon dating completed by the end of the year so as to be able to include it in their publication tentatively timed for Easter 1980.

I phoned Dinegar who, at that time, was still at the Los Alamos National Laboratory in New Mexico. He said that he had asked David Sox and Ray Rogers to serve on the committee. Dinegar said copies of his letter were sent to Harbottle (Brookhaven), Muller (Berkeley), Hall (Oxford), Fleming (University of Pennsylvania) and Oeschger (Bern). I suggested copies should also be sent to Nelson (McMaster–Simon Fraser), Schmidt (Seattle), Damon (Arizona) and Milton (Chalk River). I expressed concern that McCrone might be involved—Dinegar assured me he would not. I also said that we must operate independently of STURP. I added that it must not be a blind date.

On 3 May Seth Hill of *In Search of* arrived along with the camera crew from Channel 10. I was disappointed to learn that Lenard

Nemoy would not be putting in an appearance. I suggested to Hill that he must be getting a kick out of interviewing someone with ears whose size outrivalled Nemoy's and he replied "Yours aren't even in the same class!".

A few days later, I met at Brookhaven with Harbottle, his colleague Ray Stoenner, and Meyer Rubin. At that time, following Gonella's suggestion, we were thinking in terms of one of the Raes samples—now back in the safekeeping of the Turin authorities. We decided to request half of the larger piece he had removed from the main body of the shroud. The reported weight of that piece was about a tenth of a gram.

On 18th May I wrote to Dinegar saying that by now he should have received the material I sent to the archbishop of Turin on behalf of Rochester and Brookhaven offering to date the shroud. I gave Dinegar details of the sample sizes we would require— 25 milligrams of cloth for Rochester and 30 for Brookhaven— constituting about half of the Raes main body sample. I said the only conditions we would set would be that we would not be required to follow any blind dating procedure, that no third party should be involved in sample preparation and that we be free to publish the results independent of STURP.

Toward the end of May, I received a phone call from Sox regarding the STURP carbon-14 committee. Sox said they were worried that Rochester and Brookhaven would steal their thunder. A most remarkable statement since the Dinegar committee had no thunder to steal. He thought I ought to assure Dinegar that we realized that STURP had spent a lot of time and money on the shroud project and that surely some reasonable agreement could be reached that made them feel a part of the carbon-14 action. I had very ambiguous feelings about STURP. On the one hand they were too convinced in their hearts that the shroud was Christ's burial cloth. On the other hand they had good connections in Turin and could be useful in obtaining a shroud sample for dating—if only they could be prevented from playing any other role.

I finally decided I would call Dinegar and try to reassure him there was a role for STURP to play in the carbon dating enterprise. We discussed how the shroud sample might be transmitted from Turin to the USA. It seemed to me that Gonella would make an appropriate courier. Dinegar said that he and probably Rogers would want to observe the sample preparation and even the measurements. I did not demur at this. So far, only Rochester and

Brookhaven had responded to the letter he had sent to various AMS groups I had suggested. He wondered, if Rochester and Brookhaven did get to date the shroud, whether we intended to let the archbishop learn the results by reading a newspaper report. Also, since Rochester would get a result well before Brookhaven, would we inform them of our result? I said we would not reveal our result to Brookhaven and would also make sure that STURP and Turin were informed before going public.

Dinegar wrote me the next day thanking me for my phone call. He said he had now had a chance to read all the material I had sent him and, without being presumptuous, he thought the proposal was excellent. He said David Sox felt the same way and that Ray Rogers was considering it now. He would be surprised if Rogers came to any other conclusion. He made a number of other comments and suggestions in his letter having to do with the connection between the carbon-14 measurements and other measurements already carried out by STURP.

He realized that the carbon-14 measurements would be regarded by the general public as the most important. Clearly, he desperately wanted STURP to somehow be a part of the dating enterprise. He was, naturally, very pleased at my accepting the suggestion that his STURP C-14 Dating Committee be present during the sample taking and preparation as well as at the measurements themselves. He wondered how a pre-Easter 1980 carbon-14 date announcement would affect the *National Geographic*/STURP agreement and had asked Jackson and Jumper to comment.

He wrote that the dates determined by Rochester and BNL should not be made public independently but, rather, at the same time and place. He added he hoped I would agree that STURP and Turin would be present and 'project-connected' when the results were announced to the world. He said that, in his opinion, we were standing on the threshold of one of the most important announcements of the 20th century. He thought that several Turin trips at a minimum might be necessary and financing would be a factor. He suggested that some samples of the altar cloth that was used to patch the shroud in 1534 be used as controls because this material, connected with the shroud for so long a period of time, seemed particularly ideal.

I replied to Dinegar's letter on 31 May 1979 by saying that we had always made it perfectly clear that we and Brookhaven would make the measurements independently and without collusion. As a matter of simple courtesy, the archbishop of Turin would be informed of the results before making them public. I went on to say "We never intended to overshadow the work of STURP. That group has devoted substantial efforts to their Shroud research for many years, as well as providing substantial out-of-pocket expenses. As such, STURP and representatives of Turin should be present at any public announcement. I am not sure what you mean by 'project-connected'. I would not expect either STURP or Turin to actually collaborate on the measurements at Brookhaven or Rochester and assume that is not what you meant. I must confess, I think it a bit of an exaggeration to say that we are standing on the threshold of one of the most important announcements of the 20th century. I am reminded of the *New Yorker* cartoon showing two Egyptologists standing in the desert. They have just uncovered a four-sided triangular pointed object which looks like the top four inches of a pyramid. One says to the other, 'This may be the archaeological find of the century depending, of course, on how far down it goes'."

Dinegar's statement that announcing a carbon date before Easter 1980 might impact STURP's agreement to give the *National Geographic* sole publication rights on their shroud tests irked me. I pointed out in my letter that neither Rochester nor Brookhaven had made such an agreement with that magazine nor would we consider doing so, so why should we be bound by any such agreement STURP had made? I asked him why he thought that several trips to and from Turin would be necessary—after all, we are talking about transporting a few snippets of thread, not the chapel in which the shroud is housed! Finally I noted that there were many possible cloth samples that could be used for controls, e.g. the Egyptian bull mummy sample we had already run. "I would prefer to agree to date only the Raes sample taken from the main body of the Shroud, at least for the time being." It was readily available, it had been suggested as appropriate by Gonella and, at that time, I still assumed that there were no doubts about its integrity. Copies of Dinegar's 25 May letter and this reply were sent to members of the Brookhaven and Rochester–Toronto–General Ionex groups as well as to Meyer Rubin and David Sox.

I received a letter from Meyer dated 6 June 1979. He acknowledged receipt of Dinegar's 25 May letter and my reply. He wrote, "Leave it to the Air Force to invent STURP" and, "You say 'I am not sure what you mean by project-connected'. I think you know very well what they mean. Present and 'project-connected' was a phrase they thought about for a long time." He also noted other snippets from Dineger's letter including the suggestions that we are standing on the threshold of one of the most important announcements of the 20th century, of an Easter 1980 announcement of the shroud's age, that several Turin trips would be needed, and of using material so intimately connected with the shroud for so long a period of time. He ended by writing "Harry, I'm glad you are keeping yours, while all about you are losing theirs. There's a thing called religious fervour, but I recognize religious flakiness when I see it."

It was good to get an independent opinion that I wasn't going off half cocked about STURP.

I received a letter from Seth Hill, the writer/producer of the *In Search of* programme being prepared on the Turin Shroud, enclosing a 'location release' for me to sign which would permit them to use the portion of the programme they had shot at NSRL. He said "Your interview went very well...top quality, both visually and content-wise. It will make an excellent concluding segment in our show on 'The Shroud of Turin'. ...We're luckily getting the very top people worldwide for this show and I'm sure it will be the best in this series ever done."

In late May 1979, I received a letter from a Joe Nickell enclosing an article he published in the November/December 1978 edition of the *Humanist*. It described him as a person who studied art at the University of Kentucky and who has worked professionally as a stage magician, private detective and museologist. The article was titled, quite modestly, 'The Shroud of Turin—Solved!'. In his letter, Nickell told me that, as a result of this article, he had had an exchange of letters with one of the more fanatical members of the STURP team—a man with whom I had clashed during the 1978 meeting on the shroud in Turin—Ken Stevenson.

In the article, Nickell described how he (Nickell) had soaked a cloth in hot water, molded it to a bas relief of a man's head, allowed it to dry and then applied pigment by stroking it on with a dauber much as one would do a rubbing of a gravestone. This produced an image which he claimed was strikingly similar to a photographic negative and even superior to that of the shroud. When he brought this idea to Stevenson's attention, Stevenson quite correctly pointed out that Nickell had merely added yet another theory to the many as to how the image was formed. He then went on to accuse Nickell of not being objective. One cannot quarrel with this charge—Nickell flatly states that the shroud is a fake. I was amused, however, to hear this charge made by Stevenson—a man who believed the shroud was Christ's burial garment as passionately as Nickell believed it was not. In one of Nickell's letters to Stevenson he requested a list of names and addresses of the scientists who are involved in the (STURP) tests, and the designation of which ones Stevenson considered to be sceptical. Stevenson refused this request on the grounds that "we don't have time to be concerned about researchers who are not objective". This was a remarkable statement from someone who had already indicated that he believed the shroud was Christ's burial cloth.

In my reply to Nickell I made the obvious point that no scientific test could prove the shroud was Christ's burial cloth, but if the radiocarbon date showed it to be appreciably less than 2000 years it would be "rather good evidence that it is a fake". I said I found his explanation of the image very interesting but agreed with Stevenson that it was just another theory and not a proof that the shroud was a fake.

Dinegar wrote to me in mid-June 1979. He had received a reply from Doug Milton, head of the Nuclear Physics Branch at Atomic Energy of Canada's Chalk River Laboratory. Milton said they would not be ready for at least a year to carbon date items of historical and archaeological interest.

Dinegar noted that the letter from Milton mentioned a carbon-14 meeting in Bern, Switzerland, in August. He thought that this meeting would be the ideal place to publicly announce the plans for dating the shroud and for the transfer of the threads. He admitted this smacked of headline hunting but he also remembered reading somewhere that a "light should not be hidden and that if something is worth doing it is worth discussing from a house-top". (Surely this belongs in the category of addled adages.) He wondered who was running this meeting and whether we could get on the programme? He asked whether I or Harbottle were going to attend. He said he was posing the same question to Turin to see whether Gonella or someone would be willing to make it a threesome and hand over the threads at the meeting. He noted that Turin would not find an easier way to transfer the threads.

I was temperate in my reply. I suppose I thought that it was still worth while having amicable relations with this group of fanatical people because they clearly were in Turin's good graces. Besides, Dinegar was a genuine scientist and a friend of Harbottle, for whom I had great respect. I gave him the information he requested about the meeting and the name of its organizer, Professor Hans Oeschger, one of the giants in the field of radiocarbon dating. I said that I would be giving an invited paper on accelerator mass spectrometry. I pointed out to him that Professor Oeschger would certainly have to be consulted about any announcement concerning dating the shroud.

I attended the *IV International Conference on Ion Beam Analysis* held in Aarhus, Denmark, from 25–29 June 1979 and gave a talk on our AMS work on 27 June. On 2 July I gave a similar talk at the Niels Bohr Institute in Copenhagen, Denmark. Both Aage Bohr and Ben Mottleson attended. I had known them well since the time I spent a year at the Institute in 1961–62. Both of them, along with James Rainwater of Columbia University, had been awarded the Nobel Prize in Physics in 1975. Aage Bohr was particularly enthusiastic about the development of AMS and I had a long discussion with him that evening on the subject. He clearly would have liked the Institute's tandem accelerator group to get into the field but they never did.

After further reflection and discussions with several of my colleagues, I wrote to Dinegar that I thought it inappropriate to have any publicity connected with the shroud at the upcoming *Tenth International Radiocarbon Conference*. Although it might be an appropriate occasion to transfer a shroud sample to one of our group, no publicity should result, at least until the sample was actually safe in our laboratories. A few days later, on 25 July 1979, Dinegar wrote to me saying that the question was now moot because Oeschger had never replied to his inquiry and that it was highly unlikely that Turin would respond to his recommendations regarding providing a sample by the date of the conference. He went on to write that Ray Rogers wanted to do some textile-characterization studies on any Raes cloth released—before the carbon-14 tests were done. He thought that that made sense to everyone so he had recommended to Turin that the carbon-14 material come to me and Harbottle via Jackson/Jumper/Rogers.

Quite clearly STURP was trying to establish a foothold in the carbon dating enterprise one way or another. However, in keeping with my policy of civility toward STURP, I replied to Bob Dinegar that his proposal regarding the Raes sample seemed sensible— maybe at that time I really thought it did.

In a letter to me dated 25 July 1979, Luigi Gonella wrote that, with respect to the radiocarbon dating of the Shroud of Turin, there was an embarrassing situation, because my letter of 16 February

1979 (offering to date the shroud), never reached the hands of the archbishop of Turin. He said the archbishop authorized him to so inform me. He said he had talked to the archbishop about the carbon dating, to which the archbishop had no objections in principle, although he was a bit worried about possible polemics and interference between carbon dating and the studies following the October tests (he was referring to the measurements made by STURP *et al* on the shroud in Turin in October 1978). Gonella suggested it would be advisable for me to write another letter to the archbishop—now also a cardinal—at his official address. He included the address and went on to suggest that a copy might also be sent to him as a member of the Coordination Commission for Tests on the Holy Shroud. He said he planned to attend the *Bern Radiocarbon Symposium* where we could discuss matters at leisure. Since that meeting would take place in a couple of weeks, I decided I would hand-deliver copies of the Rochester–Brookhaven offer to date the shroud to Gonella in Bern.

The *Tenth International Radiocarbon Conference* was held between 19–26 August 1979 with the first half taking place in Bern, Switzerland, and the second in Heidelberg, Germany, some 300 km away. Although he did not register as a delegate to this meeting, Luigi Gonella attended the Heidelberg part of it. I handed him a new letter to Cardinal Ballestrero, along with the February 1979 letter and the Rochester and Brookhaven proposals. The latter had never been delivered to the cardinal. He guaranteed they would be this time and lived up to his promise. However, their receipt was only acknowledged by the cardinal some seven years later.

In late September 1979, I received a phone call from Dr F Zugibe. He was the pathologist and medical examiner in Rockland County, New York, whose hobby involved a study of crucifixion. He attended the congress on the shroud in Turin in the fall of 1978. Zugibe owns his own cross and, for all I know, suspends himself or friends from it from time to time to study the finer points of muscle strain, constrictions to breathing and other fascinating aspects of the effects of crucifixion. He had written a book on Christ's crucifixion and Rinaldi had advised him to have it published in Britain where interest in the shroud is greater than in the USA. He phoned me to get Bishop Robinson's address in the UK to get advice on publishing there. He said he had a theory from his studies of the shroud image that Christ suffered from Marfan's syndrome. Just by chance, an article on this little known affliction had appeared in the Rochester *Democrat and Chronicle* of that very day. It is an inherited disease (did God or the Virgin Mary pass it on to their son?) that affects three systems: the eye, the heart and the skeleton. In the latter, the tubular bones are excessively long, producing elongated arms and legs and spidery fingers. It is said that Abraham Lincoln suffered from it. Not all people with Marfan's syndrome have all its symptoms. What priceless information is encoded into the image on the shroud, I thought!

The above brought to mind an idea advanced by the Reverend Francis L Filas S.J. of Loyola University in Chicago. Filas claimed that on a photograph of the shroud one could actually see details of the reputed coins on the eyes of the shroud image. One of them, he stated, could be identified as a Roman coin dating to the year AD 29. This was discussed in the September issue of STURP's newsletter. The notion that the resolution of such fine details as inscriptions on Roman coins could be possible on the somewhat uneven herringbone shroud weave seemed preposterous to me. If it were true it would certainly eliminate the need for carbon dating.

I spoke to John Jackson at the US Air Force Academy in Colorado Springs in early October. He said that Ballestrero had decided to have the question of carbon dating the shroud referred to the Pontifical Academy of Sciences. Gonella would be the one to present the case to the Pontifical Academy. Jackson hoped that Gonella would deliver a shroud sample to STURP so they could carry out the tests they wanted to do. They would then hang on to the sample until a decision on carbon dating was reached by the Pontifical Academy.

He said that individual papers on the measurements made on the shroud during the 1978 tests would be published in appropriate journals without interpretation. A summary paper interpreting the results would then be prepared and submitted to some journal, perhaps *Science*. Meanwhile, *National Geographic* was going ahead full steam on their article and hoped for publication by Easter 1980. I got the impression that Jackson was relieved that there had been a delay in the carbon dating.

On 1 November 1979 I wrote to Gonella. I noted that Jackson said that the question of carbon dating the shroud was to be referred to the Pontifical Academy of Sciences and asked whether he could confirm that. I told him about my meeting with Aage Bohr last June in Copenhagen and how enthusiastic he was about our AMS work. I had told Bohr about our offer to date the shroud.

He said he was a member of the Pontifical Academy and would be glad to put in a good word for us—so we had at least one friend in court.

That same day, Dinegar wrote an interesting letter to Gonella and sent a copy to David Sox along with a hand written letter to Sox. Sox, in a letter to me dated a few days later, enclosed a copy of both. In his letter to Gonella, Dinegar strongly stated the view of STURP scientists that "dating of The Holy Shroud should be done as soon as possible". He noted there were several reasons for this. First, while STURP's measurements and their analysis of the data collected stood on their own as worthy scientific accomplishments, no interpretation of results could be made properly without an approximate date. Second, from a popular viewpoint, carbon-14 dating was **the** test on the Shroud. All the other data, no matter how convincing to STURP, were peripheral in many minds. This was not only true of the average person but also true of an appreciable number of knowledgeable scientists. He went on to say that it had been announced by Turin authorities that carbon-14 tests would be allowed and that he thought a year or two should be sufficient to overcome any obstacles or political pressure against dating. Continued delay would give rise to serious questions about the church's sincerity in the matter. In a hand written note accompanying the copy Dinegar sent to Sox, he said the letter to Gonella was written as a result of Gonella's statement at the New Mexico meeting that, because of political pressures, it might be several years before carbon-14 dating permission might be forthcoming. It was to counter the arguments Gonella said had been raised in Turin that STURP does not need carbon-14—they've proven the Shroud authentic without it!

In the Sunday 4 November 1979 issue of the *New York Times* there was an article in their 'Ideas and Trends' section under Religion titled 'Shroud of Turin Investigation Renews Debate Over Relics'. It stated that scientists studying the Shroud of Turin had recently reported that their tests neither verified nor dismissed its authenticity but that the inconclusiveness was expected to intensify interest in the shroud. Bishop Robinson, dean of the Divinity School at Cambridge University in England, was quoted as saying that believers in Christ do not need the shroud to bolster their faith, but they appreciate the fact that, if authentic, the Shroud of Turin is one of the most truly extraordinary documents of the life and redemptive mission of Christ in existence, one that should take its place alongside the New Testament.

On 11 November 1979, I wrote to Dinegar saying that Sox had sent me copies of his (Dinegar's) letter to Gonella and his note to Sox. He had told me the question of carbon dating the shroud would be referred to the Pontifical Academy. Was this true?

I noted that, in Rochester's morning newspaper (the *Democrat and Chronicle* of 19 November 1979), Thomas D'Muhala, one of the directors of STURP and president of the Nuclear Technology Corporation in Connecticut, had been quoted as saying that all members of STURP Inc. now believed, on the basis of the scientific tests, that the shroud is authentic. In that case it would appear that carbon dating was unnecessary.

Dinegar wrote to me on 18 November 1979 before he received my 11 November letter. He included a copy of a STURP report on carbon dating signed by himself, Rogers and Sox. He stated that this report, which was dated 15 July 1979, had been sent to the archbishop via Gonella in July.

The report was very laudatory of the Gove/Harbottle proposal (presumably now in the archbishop's hands having been entrusted to Gonella by me in August for hand delivery) and of Gove/Harbottle as eminent scientists with proven reputations in their fields. It recommended that the sample cut for Raes in 1973 from the bottom right hand edge of the main body of the shroud be given to Gove/Harbottle after non-destructive examination by Ray Rogers. Ten recognized experts in the field of radiocarbon dating had been asked to submit proposals for dating the shroud but only five had responded. Of these, Gove and Harbottle had said they are ready now, Hall of Oxford said he would not be ready until the latter part of 1982, Stuart Fleming of the University of Pennsylvania declined to become involved (that made sense since he had no access to carbon dating facilities capable of handling milligram samples—why Dinegar had contacted him was a mystery to me), and Milton of Chalk River, Canada, said they were not ready yet. No replies had been received from Muller of Berkeley, Oeschger of Bern, Switzerland, Nelson of Simon Fraser University in Canada, Schmidt of the University of Washington, or Damon of Arizona.

I wrote to Dinegar that I thought his committee's report was excellent. The report had recommended a press release when Turin transferred the samples and again when the tests were under way. I pointed out that it would make it difficult to work under the enormous glare of publicity that would ensue. Could the transfer and measurements be made before going to the media?

I said that I had read that the Pontifical Academy of Sciences had met in Rome on 10 November 1979. At that meeting it was reported that the pope discussed the question of what to do about Galileo, the 17th century astronomer censured by the Catholic church for stating the Earth revolved around the Sun and thus could not be the centre of the universe as the Church then taught. Now that the Church had a better understanding of astronomy, it was thought that amends might be in order. I wondered if Gonella had presented the question of dating the shroud to them.

On 24 November 1979 the programme *In Search of the Shroud of Turin* was shown on Rochester's CBS Channel 10. It contained the interview Seth Hill had done with me on 3 May. It also contained an interview with Joe Nickell in which he discussed his theory of the image formation. An article by Nickell appeared in the November issue of *Popular Photography* on the same subject. I thought the *In Search of* programme was quite well done. It was reasonably objective—not at all proceeding from the premise that this was the burial cloth of Christ, but not closing the door on that possibility either.

The same day an article on the pope and Galileo appeared in the Rochester *Democrat and Chronicle*. The headline read 'The pope apologizes to Galileo; but how to set the record straight?' The article quoted Reverend Enrico de Rovasenda, Director of the Secretariat of the Pontifical Academy as saying, "In the first place there is no question of a new trial or an appeal. All the people involved are long dead. What the holy father has in mind seems to be a mixed commission to re-examine the case, but we do not yet know who will sit on it. There were mistakes on both sides in the Galileo trial. We now know the Earth rotates on its own axis and orbits around the Sun. But Galileo was wrong, too, in believing the Sun was the centre of the universe." It occurred to me that if it took the Church almost 350 years to act on the Galileo case, we should not be too sanguine about quick action on the shroud.

In late December 1979, Gonella wrote me a long and cordial letter to which I replied a few days later. He said that he had meant to write sooner but he had been occupied with other matters. He said that he was as disappointed as I that things were not moving faster. In August he had sent the archbishop a memorandum on the two issues at hand, namely carbon dating and a clear statement about the publication of the results. He had met with the archbishop the first week of September. He said at that meeting, however, he found the 'bishop', as he called him, rather cold on carbon-14. The bishop told him that there were so many tests under way on the shroud that he felt this new test might interfere with the others and said it would be better to wait for the conclusion of tests that had already been made before initiating others. Gonella remarked that he thought this was rather strange and he had tried to question him further about it but he realized that the bishop's mind was really set. Then the bishop came up with the idea of referring the matter to the Pontifical Academy of Sciences. He thought it might be useful to have the Academy's advice on the question and he asked Gonella to prepare a memorandum to introduce the matter to the Academy. Gonella agreed to do so.

The bishop was being subjected to considerable pressure from circles connected with the Sindonology Centre of Turin— an organization run by Don Coero-Borga—and this pressure was against carbon-14 dating (this explains why Coero-Borga had never delivered to Ballestrero our original letter offering to carbon date the shroud—why on Earth had Sox suggested him as the intermediary?). The pressure was also against Gonella who, it claimed, was selling out to the Americans. The Sindonology Centre was against free publication of the results. They felt that the bishop, or even better, they themselves, ought to keep control of the results of any tests. Gonella said the bishop took a completely opposite point of view. The idea of involving the Pontifical Academy had merit. It has an international status and is a body on which Coero-Borga would have little or no influence. He said that the bishop did not think that referring the matter to the Academy would really hurt anybody because there was nothing new concerning the dating. Gonella pointed out that there was something new. Rochester and Brookhaven had officially offered to date the shroud. At that point the bishop, with a deep sigh, agreed.

He said the bishop made it clear that the scientists were quite free to publish any of the results they found. It might be courteous if they let the bishop know in advance, but he certainly wouldn't exercise any control or censorship over publication of any results on the shroud.

There were various press releases that Gonella included in this letter; one applied the word 'relic' to the shroud. Gonella said that this was a word the bishop always very carefully avoided in connection with the shroud.

He also included a translation of an article written by Baima

Bollone whom he described as 'the forensic doctor'. The article had been published in *La Stampa*, the Turin newspaper. Bollone was making tests on shreds of the shroud and was trying to convey the idea that he was the major authority on the shroud. Gonella described it in rather unkind terms. He had told Bollone that the article was full of errors and that he, Gonella, was in complete disagreement with Bollone on the whole question of carbon dating. Bollone was even protesting the handing over of these threads to Rogers for non-destructive tests and Gonella said this marked the end of his relationship with Bollone.

The article in question was essentially a diatribe against the use of carbon-14 dating on the grounds that it would not be able to solve any questions on the authenticity of the object. There was one interesting piece of information in the article, however. Bollone said that in 1961 the inventor of carbon dating, Willard Libby, became a Nobel Laureate after having dated many ancient objects. Libby then asked to test the shroud. There was a bulky set of correspondence between Libby and the Turin authorities on the question. Finally the test was not allowed because of the great amount of cloth that would be destroyed, about 20 grams. Bollone estimated that would amount to about 870 square centimetres, which turns out to be an area of cloth about 12" × 12", i.e. the size of a man's handkerchief. Bollone concluded this article in *La Stampa* by saying that it would have been reckless to allow Libby only 20 years ago to destroy uselessly 870 square centimetres of the shroud. Since then, experts have indicated that in future years it will be possible to reduce even more the amount required while increasing the precision of the answer. Therefore, why not wait until the amount becomes sufficiently small and the accuracy sufficiently high that it would be worthwhile making the measurement?

Before preparing to write the report to the Pontifical Academy, Gonella said he had consulted people on the question as to whom in the Academy it should be sent. The people who were suggested were Dirac, Weisskopf and Bohr. He said I should not worry about lobbying members of the Academy—just to go ahead and do so. Coero-Borga and Bollone would lobby as much as they can but he hoped the Academy would be out of their league.

He went on to say that he was sorry to confess that he had not yet written the report. He had been appointed to the National Advisory Commission on Nuclear Safety and he had been completely occupied writing a commission report. He would try to have the Academy document written sometime in January. One of the disadvantages that our lab, or any AMS lab for that matter, had is that there had been relatively few AMS carbon dating measurements on the record as compared to ones obtained by the classical method. He would appreciate any scientific information I could give him on our method and on the question of possible contamination of the shroud cloth. He also wanted an assessment of the accuracy of the method.

Gonella mentioned that he had a short visit with the bishop in October to report on a STURP meeting that had taken place in New Mexico and the bishop was pleased that there were no indications against the shroud's authenticity. The bishop added that we must continue with the utmost objectivity and serenity and he ended by saying "If you were to come to me at the end telling me that you discovered that the shroud is a manufacture of 50 years ago, I would accept the fact with the utmost serenity".

I had asked Gonella whether the proceedings of the 1978 congress had been published and if so could he send me a copy. He said that indeed they had been and that the paper that I presented was in the communications section at the end of the proceedings. These were papers that could not be presented orally for lack of time. He said that unfortunately our paper was in very bad company because, in these communications, there were papers that no serious congress should have accepted, they were so outlandish.

He said that in September he had phoned the rector of the University of Turin, Professor Cavallo, a biologist, who was president of the congress in 1978, and he told him about my being refused the opportunity to give a talk at the congress and about having my paper relegated to this communications section. Cavallo became very angry because he did not know anything about the arrangement of speakers. All such matters were in the charge of the scientific secretary who, interestingly enough, was Professor Baima Bollone, the 'forensic doctor'. Cavallo was appalled that our paper appeared next to a paper by a Milan radiobiologist who wrote of Jesus' auto-resurrecting by yoga techniques. He asked Gonella to tell me when he next saw me that he was very sorry, that he was just a working horse (the word 'cavallo' means horse in Italian). He had been asked to preside at the congress without having any say in its organization. He hoped that I would not be angry with him or the Turin University for the bad treatment that I received.

In my letter of reply to Gonella, in late December, I described how we establish the uncertainty in our measurements of the ratio of carbon-14 to stable carbon and I gave him a specific example. We measure this ratio in an unknown sample and compare it with a sample of known age, and we then repeat the measurements many times. This takes care of both statistical and systematic errors.

With respect to the question of whether this new accelerator method was more accurate than the conventional counting method, I referred him to the 1978 paper by Minze Stuiver in *Science*. The accelerator systems that were presently being built specifically for carbon dating are being guaranteed for accuracies of ±30 years. Most existing conventional radiocarbon laboratories do not do as well as this. I sent him a table of our most recent Rochester radiocarbon results and pointed out that we had also measured several other cosmogenic radionuclides. I felt we had accumulated a great deal of experience, even though we did not have large numbers of radiocarbon dates.

I then wrote, "Let me comment on various other points in your interesting and informative letter. First, the bishop's stand on the question of publication is excellent. Second, his suggestion of referring the matter of carbon dating the shroud to the Pontifical Academy of Sciences is also excellent. That would remove the question from parochial Turin church politics." I mentioned that I had discussed the shroud dating with Weisskopf, a member of the Academy, who had visited Rochester recently. I went on to say that I was a little concerned about how long it would take for the Academy to make a decision. They meet in Rome on average every two years, and the last time they met, which was in 1979, the dating of the shroud had not been referred to them. I remarked that the Bollone piece in *La Stampa* was certainly outrageous. Bollone seemed to conveniently forget that a piece of cloth had been permanently removed from the main body of the shroud for examination by Raes. This was a large enough piece for us and Brookhaven to date. I said that if I were the bishop, I certainly would be serene about a carbon date that said the shroud was only 50 years old. That would be a clear indication that something had gone wrong with the dating technique, because the shroud is known to be at least 600 years old. How serene would the bishop be if the age were actually established as 600 years?

Finally, I said that I was more amused than annoyed by the way in which our paper was treated in the congress proceedings.

I asked Luigi to please assure Professor Cavallo that I bore neither him nor Turin University any ill will over the matter. I asked him to send me a copy of the proceedings and he did.

The issue of *National Geographic* magazine dated June 1980 with the article on the shroud finally arrived. It was titled 'The Mystery of the Shroud'. It was written by Kenneth Weaver, senior assistant editor of that magazine and it was a really first-class piece of journalism. It had a large number of photographs showing the STURP team laying hands on the shroud when it was mounted on the famous rack that they had constructed for it. It included pictures of Jumper and Jackson trying to produce the three-dimensional model of the image. It showed Max Frei, the Swiss criminologist, along with Ray Rogers removing samples of pollen and other material from the surface of the shroud using sticky tape.

It ended by discussing the question of the age of the shroud. It noted that conventional radiocarbon dating could determine its age but it had not been permitted because of the amount of material that would have to be destroyed. It described a new accelerator technique. It quoted me as saying that with one square centimetre, the size of the tip of one's little finger, I could provide an age conservatively accurate within 150 years. It claimed that indications from Turin now led to the belief that Turin's archbishop will eventually give permission for carbon dating.

In the 15 June 1980 issue of *Applied Optics* there were several articles on STURP's investigations of the shroud carried out in 1978. These were the first actual scientific papers that were published by STURP. They contributed little to elucidating the mystery of the shroud's image.

Another scientific article by the STURP team appeared in the journal *X-Ray Spectrometry*, volume 9, No 2 (1980). None of these articles were definitive in terms of establishing one way or the other whether the shroud image was similar to a photograph or not. They mostly concluded that the evidence was against its being a painting. On the other hand, they didn't really present any indication of what else it might be.

On 22 September 1980 I discussed with Sox at the American School in London the McCrone announcement concerning the nature of the shroud image that had just hit the press. He said that Walter really was saying that the image was a fake. Sox had visited Muller in California during the summer, and was convinced that they were not close to being ready to date the shroud. In his opinion, pressure for dating was building to a fever pitch.

In order to discover what other publications STURP had in the pipeline, I phoned John Jackson and learned he had left the Air Force and had moved to Colorado State University in Fort Collins. I finally got Eric Jumper's home phone number. He said there would be a paper on the analysis of the 'blood' in the August edition of *Applied Optics*—there is definitely iron oxide in the blood areas. He said there would be three final reports: one by Ray Rogers, one by Jumper and one by some Italian author. I wondered to myself why three? He said McCrone had resigned from STURP. Driven out more likely, I thought to myself! He told me that Ray Rogers had about half of the Raes main body sample but it was unravelled! Why unravelled? God only knows—probably another of Bollone's machinations. He thought it very unlikely that Rogers would return the sample. He said McCrone was editor and publisher of *The Microscope* and would probably publish his findings there. Jumper thought it was a pity McCrone 'blabbed' before publishing. He did not think McCrone had proved the shroud a fake.

Sox wrote to me from London on 29 October 1980. He said he had been up to his ears in writing a book titled *The Image on the Shroud*. It would officially be published in January and he would get a copy to me as soon as it was ready. I did receive a copy several months later. In it he made a strong case for carbon dating the shroud, calling it the obvious test. In his letter he seemed rather pessimistic about carbon dating, given the chaotic state of affairs at the moment. He thought that the revelations from Walter McCrone that the shroud was possibly a fake would give a boost for carbon dating. However, nobody else seemed to see it that way.

On 24 November 1980 I spoke to Ray Rogers in Los Alamos. He said he and another member of STURP were trying to put together a paper on what all the STURP tests proved. Alan Adler and John Heller, both members of STURP, who are experts on blood porphorins, claimed they found them in blood areas of the shroud. They had a paper ready to send to *Science*. Rogers described Heller—a medical doctor from somewhere in New England—as a true believer. In my view Alan Adler, despite the fact he is Jewish, is one also. He is a blood chemist from Western Connecticut State College with an all-consuming interest in the shroud. Rogers claimed that the response of the image to infra-red and ultra-violet is different from the blood areas and compares well with known scorch marks on the shroud. The image is totally superficial—it is only two fibrils deep. He told me that Baima Bollone had shredded some unknown sample from the shroud and Gonella had brought 13 short threads (total weight between 40 and 60 milligrams) to Los Alamos in polythene bags. Rogers had to sign an agreement that they would not be used for carbon dating. He said he really did not know whether the threads he had were from the shroud or not. He also said he thought the side strip might not be a separate piece but part of the main body of the cloth. What appears to be two pieces of cloth stitched together was probably a tuck to provide strength for hanging the shroud.

Towards the end of 1980, Judy Brown from the University of Rochester's public relations office sent me an interesting article from *Medical World News* dated 22 December 1980. The article was headed 'The Shroud of Turin' and it had subheadlines 'medical examiners disagree' and secondly 'carbon 14 dating may solve part of the mystery'. To me one of the most interesting parts of this article was a section headed 'Will Shroud Keep Date with Carbon 14?' It mentioned Willard Libby's invention of the technique and said that he would have required a 30×30 centimetre square sample of the shroud. Then it said that in 1978 Dr Harry E Gove, Professor and Chairman of Physics at the University of Rochester in New York, officially offered to carbon date the cloth at his Nuclear Structure Research Laboratory using a sample containing a mere 6 milligrams of carbon. It said that, for confirmation, Brookhaven National Laboratory was prepared to use a somewhat different technique on a 12 milligram sample.

What was most amusing was the next paragraph. It stated that Dr Pierre Luigi Baima Bollone, Professor of Forensic Medicine at the University of Turin and Director of the International Centre of Sindonology, had told *Medical World News* he supported all requests for carbon-14 dating (had he undergone a reformation in the past year?). In fact, he was ready to do it himself if he got permission from Cardinal Ballestrero, the archbishop of Turin, and from exiled King Umberto. Bollone said one stipulation of any test was that there first be a demonstration of its capacity on a similar 2000 year old fabric, with an error rate less than 10% (as remarked previously an error of 10% in carbon dating means 830 years—anyone quoting such an error is clearly unschooled in the field). He claimed he had done such a test himself using linen from ancient tombs of the era. The Secretary of the Turin Sindonology

Centre, Don Piero Coero-Borga, who was also rector of the Church of the Sacred Sindone in Turin, claimed that carbon-14 testers had been offered but refused samples from the burned areas. Still, Bollone saw no harm in waiting since the amount of cloth needed has been decreasing. In response, Dr Gove told *Medical World News* his offer, made first to Don Coero then hand-delivered by another intermediary to Cardinal Ballestrero, had never been answered. Don Coero said Dr Gove's request may have been put off rather than turned down (a fine distinction I must say). Dr Gove denied being offered a burned sample, but said he would be happy to obtain such a piece. His laboratory has already done the required demonstration tests with less than a 1% error. Gove said "I think Don Coero is playing games".

I phoned Luigi Gonella that same day to see what progress had been made on the shroud project. He said there was no action at all. He had not even prepared anything for the Pontifical Academy (I suspected his heart was not in doing it anyhow—stemming from the Turinese fear of interference from the Vatican in 'their' shroud). He said he would be seeing the cardinal in a few days but at this time the atmosphere in Turin for carbon dating was very negative. He said it would be helpful if I were to send him a list of requests we had for dating and, even better, of dates we had already measured.

On 2 November 1981 I received a letter from the assistant editor of a journal called the *American Scientist*. They enclosed a manuscript by Bob Dinegar and Larry Schwalbe titled 'The 1978 Scientific Study of the Shroud of Turin', on the measurements made on the shroud in 1978. It was supposedly a summary of all of the findings that had been made on the shroud by STURP. The assistant editor, Jean Van Altena, commented, in her letter of transmittal, that the paper was miserably written. That was an editorial problem they could take care of, should they decide to publish. What she wanted to get from me was my judgment about its scientific content, its validity, whether proper credit was given, whether the coverage was as complete as it stood, etc.

It was a month before I got back to her. I did go through the paper carefully and on 2 December 1981 I returned it with marginal annotations on changes I thought should be made. I said in my letter that it was not clear to me that the paper contained anything new or was a thoughtful summary of the tests carried out by STURP Inc., in October 1978. I noted that the paper referenced a report titled 'Physics and Chemistry of the Shroud of Turin, a

Summary of the 1978 Investigations'. I said I hadn't seen it, but if it was published or would be published, what need was there for another review of the data?

I noted that Dinegar and Schwalbe are *bona fide* scientists, who were members of the STURP team that carried out the shroud tests in 1978. There was no reason not to publish their paper on the grounds that it might be scientifically questionable. I said it didn't strike me as the definitive review of those 1978 tests. I also noted that I was probably overreacting to the use of the phrase 'Holy Shroud' and the phrase 'The Shroud' in which both the T and the S were capitalized. I said this suggested to me something less than scientific dispassion. I ended by saying, in any case, I would mildly recommend against publication.

Chapter 4

Interlab Tests, Trondheim, Preparations for the Turin Workshop

In mid-January 1983, Michael Tite, head of the British Museum's Research Laboratory, informed Arizona, Bern/Zurich, Brookhaven, Harwell, Oxford and Rochester that the British Museum was in a position to provide two textile samples for a laboratory intercomparison test as a prelude to possibly dating the Turin Shroud. Would we be interested in participating in the programme? It was some time later, as described below, that I learned what had triggered this offer. All six laboratories accepted and were mailed two samples labelled #1, ancient Egyptian linen, and #2, cotton textile from Peru.

In early June I sent our two samples to Meyer Rubin at the US Geological Survey in Reston, Virginia to be converted to graphite for insertion in our ion source. In July, before my wife Betty and I left for Oxford on a year's sabbatical, the Rochester AMS group measured the carbon-14 and carbon-13 to carbon-12 ratio in the two samples and sent the results to the museum. The Egyptian linen sample was about 4300 years old. The Peruvian cloth was of fairly recent vintage.

In January 1984, Richard Burleigh, the person at the British Museum in charge of this intercomparison test, wrote to me. He noted that it was clear from the results received so far that the Peruvian textile sample was much more modern than they had believed it to be and another sample must be substituted. Since it was clearly important that the museum know the dates of their samples, they had found a second Peruvian sample of the age they had originally thought they had selected, but of an independently verified date. I was surprised they had not verified the age of their first sample but agreed to measure the new one. Even venerable institutions like the British Museum can err.

In Oxford I read in the London *Sunday Times* of 22 January an article entitled 'Scientists Vie to Test Shroud' written by Alison Miller. It stated that six of the world's most advanced scientific research teams, including two from Britain, are competing in a series of blindfold tests for the honour of being allowed to authenticate the Holy Shroud of Turin. It quoted Robert Dinegar, head of a committee of the American-based Shroud of Turin Research Project, as saying that delicate negotiations were under way with the Vatican through the archbishop of Turin, custodian of the shroud. Dinegar had set up a competition to see who did best in dating known samples in order to decide which of the world's experts would test a small piece of the shroud.

The article stated that the results were being collated by Dr Michael Tite, head of the British Museum's Research Laboratory. They would then be sent to the American Shroud of Turin Research Project before being offered to the Vatican for evidence of scientific capability. The scientists at Harwell were using an American innovation—a miniature counter—which enabled them to measure samples as small as 10 milligrams, and at Oxford they were using a faster, costlier invention, also American, capable of evaluating even smaller samples. Some of the results had already arrived at the British Museum's laboratory.

The article made it clear to me that Dinegar continued to fantasize about the role he and STURP were playing in carbon dating the shroud. In fact, his sole role so far was to provide the British Museum, from a list he got from me, with the names of laboratories willing to participate. The results would certainly not be sent to him despite whatever 'delicate negotiations' he claimed he had under way with the Vatican.

On 20 May 1984, just a month before we were to return to Rochester from our year at Oxford, my wife Betty died very suddenly. We had been married almost 39 years and her loss was completely unexpected and utterly wrenching. She had taken a greater interest in the development of AMS and its potential for dating the Turin Shroud than in any of the previous scientific work with which I had been involved. Now she would not share the heartaches and joys of the events that lay ahead nor provide the wise advice and support I had found so valuable in the past. It took several months back in Rochester for me to recover sufficiently to resume a more or less normal life.

In mid-November 1984, I sent our measurements of the second Peruvian sample #3 to Burleigh. It was about 550 years old. Later that month Burleigh told me our data would be incorporated in the final paper. As it turned out, we were in excellent agreement with the other laboratories except for one remarkably discordant measurement made on the Egyptian linen at one of the other laboratories.

In the spring of 1985 Easter fever again roused the interest of the press in the Turin Shroud. On 26 April a reporter from our local newspaper, the Rochester *Democrat and Chronicle*, phoned me regarding a story that I assumed had been released by the British Museum on the blind dates of the cloth it had provided to six laboratories. It appeared in the next day's paper. It noted that one group was off by about 1000 years on one of the samples and that museum officials would not say which research group came up with the wrong answer until the information was released at a scientific conference in Norway. I was quoted as saying that I was not sure what the museum officials would do with the results, but it was likely that an offer would be made to the archbishop of Turin to date the shroud cloth. Privately, I was surprised the information about the one outlier measurement had been revealed.

To track down the source of this leak, I contacted Burleigh at the British Museum on 29 April. He said the museum had not put out a news release. He did not know who did, but would try to find out. I also told him I had learned that the interlaboratory comparison paper was to be given at a poster session at Trondheim. At scientific meetings, results presented in the poster format received much less attention by conference delegates and the press than papers given verbally. Since it would likely be the most newsworthy item at the meeting, it should be presented verbally. He agreed. I said I would write to the conference organizer, Reidar Nydal, asking for this change.

I called the *Democrat and Chronicle* reporter. She said the report of the aberrant measurement appeared in a wire story carried by the *Los Angeles Times, the Washington Post* and a Long Island newspaper, *Newsday*. I suspected the person who leaked the information might be Harbottle. He headed the Brookhaven carbon dating laboratory on Long Island, one of the six involved. When I talked to him he neither confirmed nor denied that the story came from him. The discordant measurement would certainly not increase the confidence of the Turin Shroud keepers in our ability to unambiguously date the shroud. However, it would be revealed at Trondheim anyway so a premature leak would not make much difference.

He did give me more information, however, on the origins of this intercomparison test. At a 1982 archaeometry conference in Bradford, England, there were representatives from Harwell, Arizona, Oxford and Brookhaven. He couldn't remember if anyone was there from Bern or Zurich and certainly nobody from Rochester. It was Otlet from Harwell who suggested that the British Museum might be used to provide blind samples of cloth to laboratories willing to date the shroud. Hall, who was a British Museum trustee, said that he would make the arrangements by contacting Michael Tite. This confirmed what I already knew—Dinegar played no role in arranging the laboratory intercomparison.

I decided it would be helpful, in my continuing struggle to get back to the real world after Betty's death, if I were to organize a meeting of the six labs at the Trondheim Radiocarbon Conference. On 1 May I wrote to the six laboratories and the British Museum to suggest that any of their representatives who might be present at Trondheim should meet to discuss what the next steps should be.

In reply, Harbottle said he would not be going to Trondheim but that Dinegar, who was "my co-principal investigator for carbon-14 dating of the Shroud of Turin, will be there, he is certainly able to participate in this discussion which I think ought to be held as you suggest and can report the results to me". Doug Donahue responded by saying he agreed I should organize the shroud discussion at Trondheim. He said he was uneasy about the STURP involvement. I assured him that I was as well, particularly since Harbottle had suggested that Dinegar be present at this meeting.

On 17 May I informed Professor Nydal that one of the papers that had been assigned to a poster session on Monday 24 June 1985 by R Burleigh *et al* might have considerable interest to various world wide news media. The reason was the assumed relationship that it might bear to the eventual possible radiocarbon dating of the Turin Shroud. I asked him if he would, because of the importance of this paper, an importance I thought was often greatly exaggerated, consider an oral presentation of the subject at Trondheim. I added that I had written to the senior members of the six laboratories suggesting that those of us involved who would be present at Trondheim meet to discuss what steps, if any, ought to be taken with respect to the eventual dating of the shroud. I thought such a meeting might be held following the Monday or Tuesday sessions.

Nydal replied that, even before my letter reached him, he had recognized the significance of the paper by Burleigh *et al.* It had been scheduled for oral presentation on Monday, before lunch, and that information had been printed in their abstract volume. He said he would certainly find the time and place for the informal meeting I had suggested.

On 22 May Hall agreed to the meeting. He added that matters could hardly be expected to advance very much since he doubted the Catholic authorities really wanted the shroud dated. He said he could not really blame them.

In mid-June, Shirley Brignall, who had been working with me on the shroud enterprise, and I set off for Trondheim via London. On Tuesday 18 June I visited Burleigh at the British Museum. He said that his colleague, Sheridan Bowman, would be reading the British Museum paper at Trondheim although she was not a co-author. He thought that, rather than following the suggestion I made to him that the paper should have multiple authorship, the laboratory people and their institutions should be listed in an appendix. That seemed satisfactory. The British Museum paper appeared in the journal *Radiocarbon* **28** 571 (1986) titled 'An Inter-comparison of some AMS and Small Gas Counter Laboratories'. The authors were Richard Burleigh, Morven Leese and Michael Tite—all of the British Museum. Although the paper did list the six laboratories that took part in the exercise, nowhere were the names of the people involved at each of the laboratories mentioned. I considered that very unfortunate. People collaborating in any scientific measurement deserve to receive credit for their work. Perhaps, since each of the six groups involved some half dozen people or so, he thought that would be too many to list. He was clearly not familiar with some fields of physics, notably experimental particle physics, where papers co-authored by many more than three or four dozen people were not at all uncommon.

We then flew to Trondheim to attend the *12th International Radiocarbon Conference*. On Monday 24 June the British Museum paper was presented by Bowman. At the end of the talk, someone in the audience asked her what the next step would be. She said, "I'm just reading this paper, I wasn't actually involved in the work, and so I don't think I am in a position to say". At that point, Dinegar rose and said he represented the organization STURP. He claimed STURP was responsible for the intercomparison test. He noted that STURP had very close connections with the archbishop of Turin. STURP would take care of all the political arrangements connected with dating the shroud as well as that of announcing the result. He essentially advised the rest of us to wait until we got a sample and then just go ahead, like good technicians, and date it—then pass the results on to him. It struck me as a most extraordinarily arrogant statement. STURP played no role in the laboratory intercomparison and, in my judgement, had no credentials for playing any role at all in the carbon dating—if it ever took place.

The next day, Nydal arranged a meeting room for us. The representatives of the various laboratories met there and I chaired the meeting. Those present were George Bonani and Willy Woelfli from Zurich, Sheridan Bowman from the British Museum, Janice Boyd (who turned out to be Dinegar's daughter) from an organization she listed as Sycon, Shirley Brignall and Nick Conard from Rochester, Bob Dinegar from STURP, Doug Donahue and Tim Jull from Arizona, J A J Gowlett and Robert Hedges from Oxford (Hall was not present), and Bob Otlet from Harwell.

At this meeting I made two proposals. The first was that if Turin agreed to having the shroud dated, the British Museum should again assume the coordinating role it played in the laboratory intercomparison. That was questioned by Otlet. He worried that there would be two carbon dating labs involved from the UK, one his and the other Oxford, and, in the case of the Oxford lab, there was the even more disturbing fact that not only was Hall head of that laboratory, but he was also a trustee of the British Museum. He wondered whether that might look like a bit too much British influence. No one else raised any objection to the British Museum's participation, however.

The second proposal related to my feeling that it was important that we deal with Catholic authorities other than the prelates in Turin. One possibility was to somehow get the Pontifical Academy of Science involved. I said I wasn't sure what role STURP could play. Dinegar seized this opportunity to make an impassioned plea that STURP had to be intimately involved with this carbon dating enterprise. As chairman of the meeting, I found his discourse intensely annoying, especially in light of the offensive remarks he had made the previous day, but I didn't know how to silence him. He pleaded the case as eloquently as he could that STURP's long involvement with the shroud gave them a very special place in the exercise. No one else at the meeting—except possibly his daughter—seemed persuaded that his case had much merit. He was essentially ignored.

It was finally suggested that I prepare a protocol for proceeding with a dating of the shroud. It would be presented to the authorities in Turin or the Vatican after being submitted for comments or changes to the six laboratories that might be involved in the dating. When I returned to Rochester, I began to prepare this protocol.

In a letter dated 15 July 1985, sent to all the participants of the meeting held in Trondheim, I summarized the agreements and decisions reached there that led to the Trondheim Protocol, the first draft of which I included. I asked them to suggest changes or additions they would like to have made in the document. As a result, it was very thoroughly reviewed by all the people involved at that meeting. I noted that I would send a revised version of the protocol to the President of the Pontifical Academy of Sciences as soon as I had put it in a final form. The final version was dated August 1985.

In preparing this protocol, I considered the question of how one might make a blind measurement of the shroud. Although I was strongly of the opinion that every laboratory should receive a shroud sample, I did not agree with the arguments for doing the measurements blind. However, because the consensus at Trondheim was that they should be as blind as possible, I reluctantly concurred.

One question was how many control samples would be required and what should their ages be to make the results really blind? The maximum age of the shroud was around 2000 years if it were Christ's. Since its first historical mention was on the occasion of its exposition in Lirey, France in 1357, its minimum age must be about 630 years. Thus the ages of the test samples should lie between 2000 and 600 years. To minimize the time the laboratories would have to spend on the measurements, it seemed reasonable that only two control samples in addition to the shroud should be supplied to each. There would then be four possible combinations for the ages of the three samples each laboratory received, depending on the choice of test samples and on whether the shroud was 2000 years old or appreciably younger: (a) all three 2000 years old, (b) all three appreciably less than 2000 years old (but not less than 600 years), (c) two 2000 years old and one appreciably less, and (d) one 2000 years old and two appreciably less. In the first two cases the labs would immediately know whether or not it could be Christ's. The latter two combinations would not reveal the answer. So, even if the dating were done blind, there was a 50:50 chance the cat would be out of the bag. Clearly the integrity of the tests would still depend on the laboratories' agreement to refrain from premature revelation of their results.

Before the final version of the protocol had been prepared, I was contacted in late July by Dr Vittorio Canuto. Canuto was a theoretical astrophysicist working at the NASA Institute for Space Studies in New York City. He had served as a scientific aide for some time to Professor Carlos Chagas, president of the Pontifical Academy of Sciences. He assisted in organizing various conferences and meetings that the academy sponsors. He and Chagas were close friends.

He said that Chagas had recently received a 1984 proposal from STURP. It listed 26 tests they wished to make on the shroud with carbon dating listed sixth. The document, in STURP's usual fashion, was treated with great secrecy and, as a result, I have never seen a copy. It was sent to the cardinal in Turin and he gave it to the Vatican's Secretary of State who, at that time, was Cardinal Casaroli. Casaroli in turn passed it on to Chagas at the Pontifical Academy of Sciences. Chagas sent the proposal to a couple of people for review, one of them being Canuto and another a member of the Pontifical Academy of Sciences in the US. When Canuto received this STURP document from Chagas, he phoned me.

I wrote to Canuto a few days later. I included the same material I had sent to the people who attended the Trondheim meeting, noting I had only seen the carbon dating part of the 1984 STURP proposal. It was only in the call from him that I learned that carbon dating was just item 6 of 26 proposed new measurements on the shroud. Carbon dating is by far the most important test and certainly the test that should be made first. I stressed that the enclosed protocol was a preliminary version and did not reflect any changes that might be proposed by the other participants.

Canuto contacted me in mid-September saying that Chagas would be coming to New York from Boston on Saturday 12 October. He suggested 14 October as a date on which I might come to New York to talk to Chagas. He said he would send a copy of the protocol to Chagas in Brazil.

On 23 September I wrote to Professor Chagas. I said I assumed he had received, through Canuto, a document dated 10 July that I had sent to the people attending the informal meeting I organized in Trondheim. The next step was for the six laboratories to ask the British Museum to participate as the coordinating institution. It would require the approval of the museum's Board of Trustees but that was virtually assured. My idea was for the Pontifical Academy of Sciences to be the organization with which the scientists who had offered to date the shroud would interact. However, the laboratories had to proceed independently in establishing procedures to be followed. It was important that there be the clearest possible differentiation between those who offered to provide the radiocarbon measurement of the shroud and those who would decide whether or not the offer should be accepted. I ended by saying that my colleague Shirley Brignall and I at Rochester looked forward to meeting him in New York on 14 October 1985.

On 14 October Shirley and I arrived at the Holy See Mission to the United Nations in New York City for the meeting with Chagas and Canuto. They met us in a sitting room area of the mission. Chagas' English was excellent. He was a short trim man, probably in his seventies, a bit frail. Canuto was probably in his late forties, very Italian, dashing and outgoing. We learned that he had been a classmate of Luigi Gonella at the University of Turin.

Chagas described to us some of the work that his involvement with the Pontifical Academy had entailed, in particular a satellite TV discussion on nuclear disarmament between Moscow and Paris, with McGeorge Bundy on the Paris team. I remarked that MIT Professor Victor Weisskopf, a member of the Pontifical Academy and a person I have known well since I did my graduate work at MIT in the late 1940s, had been very active in appealing to the pope to take a very strong antinuclear stand. Chagas remarked, a little huffily I thought, that all the Academy members felt the same way.

He had known about the interest in dating the shroud by these new techniques well before receiving the STURP proposal. However, the idea of involving the Pontifical Academy had never occurred to STURP. Chagas gave us the credit for actually suggesting it (as indicated in the previous chapter, Cardinal Ballestrero instructed Gonella to approach the Academy but he never got around to doing so). He made it clear that he himself was not in favour of dating the shroud, but that if it must be done, it must be done in the best possible way. Despite his feelings he, as President of the Pontifical Academy, would support a scientifically reliable dating of the shroud. He had never seen the shroud. He knew nothing whatsoever about Turin politics. He said that he had met Ballestrero but that was about all. He described him as a pastoral cardinal. Canuto remarked that he and his eight year old son had seen the shroud at the time it was on display in 1978. Chagas said that he hoped that it was 2000 years old. That did not surprise me. It is what most people familiar with the shroud, including Shirley and me, hoped.

We were then given a tour of the Holy See Mission. It is a beautiful building with splendid woodwork and embossed leather panelling. There was a chapel and in a small study nearby sat an IBM word processor. That seemed somewhat anachronistic to me. In that setting, parchment and a quill pen would have been more appropriate.

Chagas said he was going to Rome in two or three days. He would try to arrange for a meeting of the laboratories some time in May, either in Turin or in Rome. He suggested that travel funds might be provided by the National Science Foundation for US participants. The whole meeting lasted for about two and a half hours. Both Shirley and I came away from it with a very good impression of both Chagas and Canuto.

In mid-November 1985, Canuto told me that Chagas had called him from Brazil that day—there was a green light to go ahead with a meeting to discuss procedures for dating the shroud. It would be held in Turin on 9, 10 and 11 June 1986. Ballestrero had approved it. I was delighted to hear that and amazed it had happened so fast—Chagas must have substantial influence with both Turin and the Vatican. Chagas felt that it was very important that the British Museum be involved. He said that I should contact the US people and I should also discuss possible financing of the travel costs of one person from each of the US labs from the NSF. The person to contact there was a Mr H Uznansky, who headed the NSF's Cooperative Science Program.

Uznansky must have been alerted by Canuto because he called me the next day to say that the NSF Program would be interested in supporting a small workshop on procedures for carbon dating the shroud. The support would include costs of international travel and modest living expenses in Rome for three or four US scientists, including me. He said he would send me the NSF guidelines for such a proposal. The proposal would be subject to peer review by outside scientists following general NSF practice.

Shortly afterward, Tite informed me that the trustees of the British Museum had been consulted—they had agreed to my request that the museum should act as the coordinating institution. He said he looked forward to hearing further from me.

On 26 November 1985 I reported the recent events to the laboratories. Chagas had obtained agreement from the appropriate Vatican and Turin authorities for a workshop in Turin starting 9 June 1986 to discuss the procedures to be followed for dating the shroud. I described who Canuto was and the role he was playing. Because of the agreement in Trondheim that the Pontifical Academy be as fully involved as possible, I sent Canuto my memorandum and our suggested dating protocol for him to transmit to Chagas. Chagas had then asked Canuto and another person in the US (a member of the Pontifical Academy) to review both our and the STURP proposal. Both reviewers favoured ours as did Chagas, a fact he made known to the Vatican. I recounted the meeting with Chagas and Canuto in New York on 14 October 1985 and how Chagas seemed to be particularly pleased the Trondheim group had emphasized the importance of the involvement of the Pontifical Academy. I said I would shortly apply to the National Science Foundation for travel and living expenses for the US representatives to the workshop. The European participants would have to find their own funds.

On the copy of this letter to Dinegar, I noted that I might have some problems in getting NSF funding for his travel. Would he be able to find some other source? This was distinctly tongue in cheek because I had no intention of asking the NSF to include him in the grant. I knew that he would find money to get there one way or another.

By this time, I was finding Dinegar's actions on behalf of STURP intensely annoying. I felt, naively as it turned out, that I now had a golden opportunity to eliminate STURP once and for all. On 6 December 1985 I wrote to Dinegar saying that I had been trying to rationalize a role that STURP might now play in carbon dating the shroud. I noted that my initial patent reluctance to have STURP further involved in any way had been somewhat modified. I realized that STURP had an obvious interest in the shroud because of the years of effort put into its investigation. Their contacts in Turin might be important in obtaining samples. Now, however, the need for STURP's good offices no longer existed. Professor Chagas, President of the Pontifical Academy, had obtained the agreement of Cardinal Ballestrero to hold a workshop in Turin to discuss details of the procedures to be followed in dating the shroud. This workshop might include obtaining samples. The people directly involved in the measurements would be there along with the appropriate Vatican and Turin officials. STURP's presence at the workshop and in all consequent activities could only be as an interested bystander. There were clearly many other groups that would like to play this role as well. I added that I was afraid that STURP had been accidentally and unwittingly pre-empted in this whole operation. I would be hard pressed to explain why STURP should be involved any further in the carbon dating process, particularly since some people already had reservations about STURP's impartiality concerning the nature of the shroud. I ended the letter by saying it would be best for STURP to wait until the shroud's age had been established. They could then renew their request to perform the other 25 tests, whatever they might be.

I was now beginning to get replies from the letters of invitation sent to the various people to participate in this dating. Woelfli agreed to attend. Hall wrote saying this was the first time anything had moved forward. He said that he and Hedges would attend. He had not been able to be present when the Trustees of the British Museum met on the question of coordinating the carbon dating but understood things had gone well. He ended by noting that, although it sounded a step forward, we were still some way off getting a piece of the shroud. He indicated a certain lack of confidence in the clerical involvement. It seemed to me, however, that would be difficult to avoid.

I also got a reply from Dinegar, dated 12 December 1985. He wrote that my letter of 6 December came as a shock, but not a surprise and that STURP completely rejected the conclusions it stated. He strongly suggested that, in the interests of individual and collective cooperation and success in carbon dating the Shroud of Turin, I reconsidered and returned to the recommendations of the protocol I submitted and with which STURP agreed. The protocol, written before the contact had been made with the Pontifical Academy, had indeed suggested STURP play the role of an intermediary with Turin.

Harbottle phoned on 13 December 1985 saying that he had been in contact with Dinegar and that he had a copy of the above letter. He said that he would write to Dinegar. He did so on the 23rd, arguing that he (Harbottle) would be a completely appropriate representative not only of one of the carbon dating laboratories, but also of STURP. There was no need for Dinegar to get so disturbed about being omitted from the list of invitees to the workshop. He suggested that STURP should be more interested in having a reliable carbon-14 date from the consortium than in claiming more than its fair share of the credit. He noted that, so far, STURP had done all the experiments, written all the scientific articles, been on all the talk shows and starred in all the films on the Turin Shroud. In his view that did not alter the relative stances that have now evolved owing partly to the fact that STURP diluted the relative importance of the carbon-14 experiments by imbedding it as number 6 out of 26 in the proposal of 1984. He said that very much disappointed him. He said that it was fairly obvious that the June workshop in Turin was under the auspices of the Pontifical Academy and that surely they can invite whomsoever they please. If Dinegar did not feel that Harbottle could competently look out for STURP's interests as well as Brookhaven's, then he should write to Professor Chagas asking to attend as an observer or in whatever capacity Chagas and he may agree on. He emphasized that it was Chagas' workshop, not STURP's, the laboratories', Gove's or the British Museum's. He ended by wishing Dinegar a Merry Christmas. It was an extremely strong letter and it is doubtful whether the dour Robert Dinegar's Christmas was very merry after being dealt such double body blows.

At the end of December, Canuto told me that Chagas had informed him Ballestrero was still dragging his feet—the June meeting was not certain after all. He said that Chagas was going to send the Chancellor of the Pontifical Academy of Sciences in Rome, Father Rovasenda, to Turin to talk to Ballestrero. Chagas was very upset about Ballestrero's lack of response.

On 2 January 1986 I submitted a proposal to the National Science Foundation's Cooperative Science Program with Western Europe. It was titled 'Workshop on Radiocarbon Dating the Shroud of Turin'. I requested a total of $5670 to cover the airfare and living expenses for the three US participants. The technical abstract noted that the workshop would settle details of the procedures to be involved in establishing a radiocarbon date for the shroud of Turin, that new carbon-14 measurement techniques would be used requiring very small samples of cloth, that six laboratories, three in the USA and three in Western

Europe, would make the measurements and that the coordinating institution would be the British Museum. The names of the three US participants were Donahue, Harbottle and me. It also listed the European scientific participants and stated there would also be delegates from the Pontifical Academy of Sciences, the Vatican and Turin.

On 2 January 1986, Otlet informed me that a recent UK government edict decreed that Harwell must become commercially independent. This would make it difficult for him to obtain travel funds to attend the Turin workshop. He said he was very keen to participate and thought that recent developments were very exciting. I told him he just had to attend the workshop.

In mid-January, Robert Hedges wrote that, while he would like to come to the workshop, he really doubted he would have the time to spare discussing for three days questions that had already been ventilated for several years. He appreciated that any move on the part of the Vatican would need to be preceded by a discussion and he would be happy to represent the Oxford lab for a day if this proved necessary. He would certainly wish to be present if ever the sampling was undertaken. He said Hall might be able to attend but doubted if he would have the time suggested available. Hedges had always been negative about radiocarbon dating the shroud on the grounds that it was not of sufficient scientific interest.

On 16 January 1986, Chagas phoned me from his Vatican office. He said that he had met with Ballestrero at the Vatican, that a meeting in Turin 9 to 11 June 1986 had been approved and that samples might even be taken at that time. He said that Ballestrero wanted to get the whole thing over with in Turin in as short a time as possible. There was a reaction against the enterprise by others in Turin and he wanted to minimize the agony. Chagas said he would send invitations to participate in the workshop. He also said that Luigi Gonella would come to Rochester. Gonella would be meeting with Chagas in Rome and Chagas would phone me tomorrow as to when Gonella was to arrive. He emphasized that there was to be no publicity whatsoever about the Turin workshop.

I contacted Tite on the 17th of January and passed on Chagas' message. Tite felt there should be a textile expert present, if samples were to be taken, to make sure that we were getting a piece of cloth from the main body of the shroud on which the image was imprinted and not a rewoven area or a patch. He said he would try to suggest the name of an appropriate person to serve as such an expert.

I called Teddy Hall at home to warn him about publicity. He said that he was concerned about the Zurich involvement because there were three cases he knew of, one involving the British Museum cloth samples, in which they had made mistakes. I told him that Woelfli was very embarrassed about the outlier incident connected with the British Museum samples. It was actually an error made by his collaborators in Bern in preparing the carbon from the British Museum cloth samples. Woelfli said that the work involving the shroud would be done directly by his group and that he was very keen to participate in this exercise. I felt very strongly he just had to be included despite any misgivings anyone might have.

When I contacted Otlet he was very excited that this adventure was under way. He said he had gotten Harwell's approval for travel and that he would refer any press inquiries to me.

I called Woelfli at home and I gave him all the above information, including the warning on press contacts and the suggestion about having a textile expert present, and he agreed with both. He was very pleased with the progress that was being made.

I also contacted Uznansky at the Division of International Programs at the National Science Foundation, warning him about publicity. He said he wanted the proposal title to be worded in such a way that he would not get Senator Proxmire's Golden Fleece Award. I read him the title and he seemed to be satisfied with it. He said that he still had some lingering concerns about the religious involvement of the project. As was the case with all NSF proposals, it would be sent to outside reviewers and he asked me to suggest a couple of names.

That same day, Chagas again phoned me from the Vatican. He said that Gonella would visit before the end of February and would contact me. Chagas said there were problems with STURP. They have many supporters in Turin. He wondered whether three days might be too short a time to establish a protocol for carbon dating the shroud and for taking samples. Apparently Gonella thought it might be.

Canuto also phoned me that day. I had sent him a copy of the letter I had sent Dinegar. He said he thought that Gonella was coming to the USA on STURP's behalf. He wondered why Gonella was involved at all. I said that I assumed it was because Gonella was Ballestrero's scientific advisor on matters concerning the shroud, but Canuto did not agree—he said he was very suspicious. I read Canuto the letter Harbottle had sent Dinegar. If Chagas thought that Dinegar should attend the workshop it was fine with me. Canuto said he had discussed with Chagas my reluctance to involve STURP because he thought it was politically unwise. He said Chagas would be calling him tomorrow regarding STURP involvement.   Canuto said he would advise Chagas to invite Dinegar and that he had consulted me and I agreed. He suggested that Gonella meet with me at the Holy See Mission in New York and that the meeting should also include him.

On 20 January I talked to Harbottle and mentioned the question of the textile expert. He said that he was sure that there were Italian textile experts who could serve. He had heard from Dinegar that he would be in Peking on his honeymoon during the 9–11 June workshop. Dinegar told him that he had been assured by Gonella that he could attend the meeting. Dinegar furthermore said that Gonella would be chairing the meeting. Harbottle said that Dinegar would cut short his honeymoon to go to Turin!! It did not sound to me like a marriage that had a long-term future.

That same day, Canuto told me that Chagas approved of the meeting with Gonella being held at the Holy See Mission. Canuto suggested to Chagas that he hold off on inviting STURP people until after Gonella's visit. He assured me that Chagas would be very tough about who he would or would not allow to attend the meeting. He once sent a US general home from a meeting in the Vatican concerning the USA's Star Wars programme, a person who had not been invited to the meeting. Canuto also said that there had never been a meeting sponsored by the Academy of Sciences that was chaired by anyone other than a member of the Academy so Dinegar's statement that Gonella would chair the Turin workshop was nonsense.

After talking to Canuto, I phoned Hedges again. He said that although he had written that he did not have time to go to the Turin workshop, he had heard that samples might be taken, so had changed his mind. He would write another letter saying he would attend.

I then wrote a long letter to Chagas enclosing a copy of the letter that I wrote to Dinegar on 6 December 1985, when it was clear that Chagas and the Pontifical Academy would be playing a role in the dating of the shroud. I assumed this was the letter that had caused STURP such concern. I also enclosed a copy of the letter Harbottle had sent to Dinegar, which Harbottle had given me permission to do.

I said that unless STURP could show that it had some role to play, I personally could not justify their presence at the workshop. I listed some reasons why I thought STURP representatives, particularly Dinegar, should not be invited to participate. First, many of us were appalled at Dinegar's statement at Trondheim, that STURP would take care of the politics and, presumably, the kudos, and that the rest of us could just go about doing our science. Second, many of us had reservations about STURP's impartiality regarding the genuineness of the shroud. Third, in STURP's proposal, radiocarbon dating was only one of 26 measurements they wanted to make and it was not even listed as number one at that. Fourth, the only member of STURP that had any carbon dating expertise was Harbottle, and he would be present at the workshop. Fifth, the frequent dissensions within the STURP organization made them appear to be a rather unstable group. However, if Chagas felt that it would be politically wise to involve Dinegar, then I would drop my opposition. I said I had faith in his judgement as to who should be invited.

I mentioned that Tite had suggested there be a textile expert present at sample removal time. I felt that the person chosen must be an acknowledged expert in textiles and not have any preconceptions about the shroud's origin. Tite would probably suggest the names of people that might be appropriate. Chagas might know of some textile experts in Italy or Brazil who would fit these qualifications.

On 10 February 1986, Gonella phoned concerning his upcoming visit to New York. He said that he would be coming as a representative of the archbishop of Turin. He would be interested in discussing what would be talked about at the workshop and about the possibility of taking samples during the June meeting. He said that Chagas had suggested that Canuto also be involved in the meeting. He stressed the fact that he was very much in favour of carbon-14 dating of the shroud. During this phone call he gave no hint at all of any STURP involvement as the reason for his visit. I said that the Vatican ambassador to the UN had kindly agreed to let us use the Holy See Mission for our meeting in New York on the 17th of February at 2:00 pm. I gave him the address and the phone number of the mission. He told me that he had lots of interesting things to tell me: for example, the whole enterprise was on the brink of collapse. I thought that was pretty good for starters.

On the 14th of February, I talked to Canuto. He told me that Chagas did not really think that carbon-14 dating was necessary to prove Christ's existence. I had always thought that went without saying. In a letter of 12 July 1985 from Chagas to Canuto, Chagas said that Ballestrero, for many years, had been very much influenced by Gonella on matters of science. He had always wanted to date the shroud and had some doubts about the scientific merits of STURP. Chagas claimed that the point of Gonella's visit to him when Chagas was in Rome was to express Gonella's concern about which textile expert would be chosen. Canuto suggested that I call Chagas in Brazil.

I did so immediately after this conversation with Canuto. Chagas made it clear that he would preside at this Turin workshop. He said that the new president of the Brazilian Academy of Sciences was about to take office and that he would consult him about a textile expert. Chagas wanted to internationalize the whole affair as much as possible. He said that it was necessary to be very tactful with representatives in Turin, since there were many people in Turin who opposed the dating. He thought that the cloth removal could take place on the second day of this workshop.

Meanwhile I had been discussing with Ted Litherland a meeting of the Royal Society to be held in London in late June 1986. He, Ken Allen of the Oxford Nuclear Physics Laboratory and Teddy Hall were organizing it. Of the three, Ted was the only member of the Royal Society. It was decided that I should give the paper on the Rochester work at this meeting. The title of the paper was 'Ultra-sensitive Mass Spectrometry with Accelerators'.

On 17 February 1986, Shirley Brignall and I went to the Holy See Mission prior to our meeting with Gonella and had lunch there with the Vatican's ambassador to the UN, Monsignor Celli. We had learned prior to our arrival that Canuto was so stricken by flu he would be unable to attend any of the day's activities. The meeting that afternoon would include only Gonella, Shirley and me.

Gonella arrived at the Holy See Mission at about 2:00 pm. He carried with him a message from the cardinal. It was addressed to me and I quote here the English translation provided by Gonella. "Professor Gove, I am glad to introduce you to Professor Gonella of the Turin Polytechnic, as my trusted representative for the matter of the C-14 test of the cloth of the Shroud of Turin under the patronage of the Pontifical Academy of Sciences. Thank you for your availability. Yours sincerely, Anastasio Cardinal Ballestrero, Archbishop of Turin." I hardly needed this introduction but I suppose Gonella felt it gave him some extra status at this rather strange meeting the purpose of which I was still uncertain.

Gonella began by apologizing on behalf of the cardinal for not replying to the Rochester–Brookhaven offer to date the shroud. He said that the cardinal was unable to say yes or no and so it seemed best to say nothing. I refrained from remarking that it would have been courteous for Ballestrero to at least acknowledge receipt—but who was I to teach courtesy to cardinals?

Gonella also informed us that he was involved in some way with the safety of nuclear power plants and that there was some analogies with the shroud, i.e. both have a fascination for nuts. His chief concern was the conservation of the shroud, which he believed to be a major issue. He then noted that the shroud was not of as great interest to the masses as other religious artifacts, but of all the existing artifacts, it is the one that incites the most interest on the part of scientists. He didn't mention what other religious artifacts were of greater interest to the 'masses' and I would be hard pressed to guess which ones they were. I doubt there are any.

He remarked that Turin would be responsible for the shroud's safety, including responsibility for taking samples of the shroud. He said that Turin would never ask for a blind test of the shroud. That surprised me since practically everyone else except me thought it was absolutely essential that the measurements be made blind.

I asked him about a couple of people who kept cropping up as ones who must attend this meeting; one of the names was Drusik. Gonella said he was a museum conservator with the J P Getty Museum in Los Angeles, and was an expert on textiles. I also mentioned the name of Alan Adler, a professor of chemistry at Western Connecticut College. Gonella said he was an expert on blood.

We then went on to discuss the question of where the samples might be taken and talked a little bit about the cloth that still remains under the patches. In fact, he brought with him and gave me a photograph of the shroud as it appears when lit from the back. One can see under the patches some relatively clean areas of cloth that might serve as samples. These pieces of cloth, however, are undoubtedly scorched to some degree and Gonella said that from the point of view of conservation it would be a good thing to remove the scorched material. If it could be used for dating the shroud, it would be good for the shroud. He used some sort of argument that charred or scorched material might migrate into unscorched adjacent areas by some process he did not describe.

Gonella said that Monsignor Cottino had died a year or two ago. Cottino was the man who appeared beside me on the steps of the cathedral in Turin in 1978 when the CBC was there taking photographs for their television programme *Man Alive*, and who had angrily rejected the idea of carbon dating the shroud. He had been the day to day custodian of the shroud and he was a very tough man. His death would remove one obstacle to dating the shroud. Many, however, remained.

Gonella then said that he had first heard of the 9–11 June Turin workshop in a copy of a letter I had sent to Dinegar concerning it. Dinegar had passed it on to the director of STURP, a man named Thomas F D'Muhala of the Nuclear Technology Corporation in Connecticut, and D'Muhala had sent it to Gonella. Since this date had been arranged between Chagas and Ballestrero, I found it incredible that the cardinal had not informed the person he had designated as his science advisor on carbon dating the shroud.

Gonella argued forcefully for the meeting to be held at the Turin Polytechnic, where he had a faculty position. His argument was that if a group of foreign scientists visited Turin, they would not attract as much attention if they were meeting at the Technical Institute as they might if they were meeting in some building owned by the church. He said he would prefer a meeting in December—June was too soon but he would accept it.

He pointed out that the shroud itself resided in a chapel, attached to but not of the Turin Cathedral proper. The chapel was formerly the property of the House of Savoy, but it now belongs to the state. The people in Turin continue to be fearful about the possibility that the shroud might be moved to the Vatican.

He said it was important, in relation to conserving the shroud, to determine whether the image was fading, and so on. He discussed the 26 proposals STURP had made for further tests. These were being proposed because new techniques had been available since the 1978 testing. He said that the STURP proposal for carbon dating (item 6 of the 26 proposed tests) was identical to the Trondheim Protocol except for the involvement of the British Museum. The STURP proposal was the only one Turin was considering. He said fortunately the question of involving the famous Turin Egyptian Museum as the coordinating institution had not been raised in the STURP proposal, otherwise this might give Turin serious problems about accepting the British Museum in this role. They presently did accept the British Museum's participation.

He said the shroud was Turin's responsibility and they believed STURP textile experts would be the best choice. He stressed that STURP must be involved in any workshop. I raised the question of how impartial the STURP scientists could be since I put them in the category of true believers. He felt that it was possible to be a good scientist and a devout Catholic at the same time. I should not judge Jackson by the cross he wore while testing the shroud, but rather on the scientific integrity of the papers published in reputable scientific journals that Jackson co-authored. I noted that, in my mind, there was a distinction between being a devout Catholic and having a passionate belief that the Turin Shroud was Christ's burial cloth. He argued that, since he assumed I was an agnostic, I might also be accused of being biased. I tried to convince him that, unlike the members of STURP, I had no preconceived notions about the age of the shroud, but I don't think I succeeded.

Gonella pressed very hard on the question of why six labs were needed, whether the accelerator laboratories could use less material than I was suggesting and whether we could use scorched or partly scorched material. The latter is consistent with his professed concern about the conservation of the shroud. I said six labs were not needed but six had offered to participate. Six had engaged in the British Museum intercomparison and the one outlier result had been detected because more than two or three labs had made measurements. The amount of material required and whether scorched material was suitable could be settled at the workshop.

It was a tough meeting that must have lasted three hours. Gonella was a very intense man despite his attempt to disguise it by puffing calmly on his pipe. Although his English was very good, he had a pronounced Italian accent that required one to listen very intently to what he was saying. Both Shirley and I were exhausted by the time it was over. Following Chagas' suggestion that we be frank and candid with Gonella, I gave him a copy of the list of proposed subjects to be discussed at the workshop.

The next day, back in Rochester, I called Canuto. He was devastated that he had not been able to attend this meeting with Gonella. He said that the workshop should be held in the Vatican, it was after all an Academy meeting and that all the Pontifical Academy meetings are traditionally held in their headquarters in the Vatican. He said that Gonella had met for an hour with Ambassador Celli after we left and that Celli had reported to Canuto that Gonella was really a fine fellow and he seemed convinced by everything that Gonella said. Apparently Gonella went to Connecticut to see D'Muhala, the STURP boss, immediately after this meeting. Canuto also said that, as far as he knew, the Italian press knew all about the Turin workshop, so secrecy was completely impossible. He said I should write a report immediately for Chagas and that I should call Chagas tomorrow. What really upset him, as much as it did me, was Gonella's suggestion that the Turin workshop should be held at the Turin Polytechnic. He said that that idea was absolutely preposterous. It would violate Pontifical Academy tradition and it would give the advantage to STURP.

Ted Litherland phoned shortly after the meeting with Gonella. He asked what steps, if any, I had taken for a tenth anniversary celebration of the AMS measurement made on 14 May 1977 in Rochester of radiocarbon from natural samples. I said none so far, but it got me to thinking about what we should do. This culminated in the holding of the *Fourth International Symposium on Accelerator Mass Spectrometry* at Niagara-on-the-Lake in Ontario, Canada, 27–30 April 1987. He also said that Rich Muller had become convinced that some major celestial body he (Muller) dubbed Nemesis periodically swung near the Earth and caused major calamities such as the death of the dinosaurs. He was interested in exploring how this would affect astrological predictions! That sounded like a good project for Muller—one that should keep him from further complicating the project to carbon date the Turin Shroud. He also mentioned that the chairman of the Fine Arts Department at the University of Western Ontario, William Dale, was an art historian whose area of interest was early Byzantine. Ted said Dale had given a lot of thought to the shroud and the possibility that it could be a fake.

On 19 February I phoned Chagas and told him, among other things, that Gonella wanted the workshop held in his institution— the Turin Polytechnic. I said I had given Gonella a draft of the possible items for discussion at the Turin workshop. Chagas' instant reaction to the Turin Polytechnic locale was that the workshop would be held in Rome and that a second trip would be taken to Turin for samples. He said that he would call Canuto and would contact Ballestrero.

He talked to Canuto that same afternoon. Chagas said that he would phone the chancellor of the Pontifical Academy, Father Enrico Rovasenda, and ask him to contact Ballestrero to say that the programmatic workshop would be held in Rome rather than in Turin. Monsignor Celli, the Vatican ambassador to the UN, would telex the undersecretary of state, Cardinal Martinus Somolo, the person in the Vatican to whom Celli reports, that the workshop would be held in Rome. I should inform Tite of the Rome locale. In this phone call, Canuto told Chagas that, according to Gonella, the Pontifical Academy of Sciences would be allowed to serve coffee at the meeting at the Polytechnic in Turin but that would be the extent of their participation. That was a pretty fair depiction of the role Gonella would have liked the Pontifical Academy of Sciences to play.

I received a letter dated 17 February 1986 from Tite informing me that he had written to Madame M Flury-Lemberg (head of the Textile Workshop Abegg-Stiftung in Bern, Switzerland) asking whether she would be willing to examine the fabric of the shroud prior to taking a sample for radiocarbon dating. She expressed her willingness to assist. He suggested that I write to her giving her further background. He enclosed his letter of invitation to her and her affirmative reply. This caused me some concern because I had merely asked Tite to suggest the name of such an expert to me. The suggestion would then be passed on to Chagas.

On 22 February I spoke to Tite. He said he thought that I had said in my 17 January phone call that he should contact such an expert. I replied that I had just wanted suggestions that I would transmit to Chagas. I said I had written to Chagas saying that Tite would make suggestions and that possibly Chagas might know some textile expert in Italy or Brazil who would be suitable. I told Tite that Chagas was consulting the President of the Brazilian Academy of Sciences for recommendations. Tite said "Why on Earth Brazil?" I told him that I supposed it was because Chagas was a Brazilian. I also told him that Gonella had suggested that a STURP textile expert would be best and that appalled me. In any case we could not be committed to Madame Flury-Lemberg. He said that he had consulted Hall and decided to ask her although he did not know her himself. Tite suggested that it would be best if I were to write to her saying that things were still up in the air. I said I would.

On 25 February Harbottle phoned from Brookhaven. William

Meacham, an archaeologist from Hong Kong, had approached an organization called ASSIST (Association of Scholars International for the Shroud of Turin, of which a Paul Maloney was the director) concerning his objection to the use of charred material for dating the shroud cloth. He thought that during the high temperature of the charring there might be some complex carbon exchange with the carbon dioxide in the air. Harbottle had prepared a letter asking various laboratories what experience they had had with charred material to see whether charring actually did affect the date.

He said Gonella visited him after our meeting in New York. Gonella told him that the Raes samples looked very suspicious and did not appear to match the cut-away portion of the shroud. Gonella did spend several days with D'Muhala, the director of STURP, in Connecticut. About two weeks after the planned Turin workshop, STURP would perform a series of tests on the shroud. They were presently gearing up for these, including more photographs, X-rays and ultraviolet measurements, etc; presumably some or all of the proposed 26 tests. They would pack all the equipment up and go to Turin as they did in 1978. It would be during these measurements that samples will be taken. I told him that Gonella had not mentioned to us in our New York meeting any imminent STURP operation.

I gave Canuto the above information, emphasizing particularly the proposed STURP measurements on the shroud two weeks after the workshop. I said I was opposed to sample taking at that time but, whenever it was done, at least the British Museum and a textile expert needed to be present. He said that he would phone Chagas that day. I then composed a letter to Madame Flury-Lemberg informing her as diplomatically as I could that, although Tite had invited her to be the textile expert, this had not yet been agreed to by Professor Chagas.

In late February, Canuto phoned to say that Chagas had just sent a telex to the Cardinal Secretary of State in the Vatican and a letter to Cardinal Ballestrero that the meeting would be held in Rome. Chagas agreed that the laboratories must be present when samples were taken and that should not occur at any putative STURP tests held after the workshop. In any case, Gonella should have informed him about any STURP tests. He also indicated that he would approve the choice of Madame Flury-Lemberg as the textile expert.

On 18 March 1986, Gonella called to discover whether I had

heard anything definite from Chagas about the workshop. I said I had not. So far as I knew it was still scheduled for the June dates at the Vatican. He reminded me that when we met in New York the meeting was to be in Turin. I said that since that meeting, Chagas had decided on the Vatican. Canuto had told me that Chagas had informed Gonella of this directly, so Gonella knew about it despite professing otherwise in this phone call. Gonella said that no word of this had come to Turin. That was strange in view of the letter that had been sent to the archbishop. I could only think Gonella was being less than candid.

I asked about the STURP tests, why had he not told me that they were definite for around 23 June? He suddenly got very vague. He said that he was waiting word from Chagas—as if Chagas were in any way involved in these STURP tests. He said that the date was not fixed. The STURP people would not be ready, and that it would probably be done sometime this year. In any case it did not involve carbon dating. I said I knew that, but what about sample taking? He made no comment and ended by saying that he would try to contact Chagas directly.

The next day, Canuto told me Chagas had called him. Chagas had been ill for a few weeks. Chagas said the meeting would now be held in Turin at the Polytechnic. Ballestrero had pulled his weight with the Vatican, I assumed at Gonella's urging. Chagas was apparently under very intense pressure and was no match for a prince of the Church, at least this was the way Canuto described the situation. Somehow or other, Gonella had passed the word to Chagas that I gave him the impression at the New York meeting that I was running the show and even worse I was an agnostic! Gonella said that it was essential that the meeting be held at his institute, not on Ballestrero territory because of the press. Canuto again expressed great opposition to holding the meeting at the Turin Polytechnic.

I indicated to him that I was desperately disappointed in Chagas' lack of leadership. It belied everything I had expected after our meeting in New York. If the Pontifical Academy of Sciences had to knuckle under to 'Princes of the Church' in matters of science, then I really must reconsider the involvement of our group. I had found that dealing with prelates, no matter how elevated, was a hopeless enterprise. I was not opposed to the workshop being held in Turin. However, it must not be held in Gonella's home territory, it must be held in a place controlled by the Turin Archdiocese. As far as I was concerned, Gonella's location was a STURP location.

That same day, I phoned Chagas in Rio de Janeiro and said I had talked to Canuto yesterday and he said that Chagas had changed his mind about the location of the workshop again. It would now be in Turin. Chagas said he hadn't changed his mind: he had problems with Cardinal Ballestrero. He did not know yet whether it would be in Turin. He had written to the Swiss textile expert and invited her to attend. He was expecting a phone call today or tomorrow with respect to the location. I said I could accept the meeting being in Turin, but I was opposed to the Technical Institute locale. He agreed. He said he was going to chair and direct the workshop. The only thing to be discussed was carbon-14, not the other things that Gonella spoke about. He said he would let me know the locale tomorrow or Monday.

By this time, I was becoming exasperated with the unnecessary complications that seemed to be involved in dating the shroud. On 27 March 1986 I composed a letter to Professor Chagas which was sent to him in Rio de Janeiro by Federal Express. I said that I was getting the strong impression that authorities in Turin, including Cardinal Ballestrero, thought people in the six radiocarbon dating laboratories and the British Museum had some overpowering desire to date the shroud. Consequently the Church thought it would be doing us an enormous favour by even grudgingly allowing us to do so.

I urged him to remind the cardinal, the undersecretary of state, and any other high ecclesiastic authorities involved, that we offered to establish the age of the shroud cloth only if the Church wished us to do so. It did not have a high rating on any of our scientific priority lists. If Turin or the Vatican thought that it should not be done for whatever reason, then we should be informed. However, if the Vatican wanted us to proceed, then substantially greater cooperation was required. We did not offer to do it for a fee and thus had a right to fully participate in decisions on the circumstances under which it was to be done. If it were decided that the workshop must be held in Turin, none of us would object as long as the workshop takes place in meeting rooms that are part of the Turin Archdiocese—not in the Technical Institute of Turin.

I added that I was very pleased he had invited Flury-Lemberg to be the textile expert involved. I concluded that we expected Chagas to run the operation, not the cardinal, nor his science advisor, Professor Gonella or STURP. I emphasized that it was the involvement of the Pontifical Academy that made the whole operation even remotely viable. Without that I suspected few of us would continue to be involved.

On 1 April 1986 (an appropriate date, perhaps) I got a very disturbing phone call from Canuto. He started by saying it never rained but it poured. He had just received from Chagas a copy of a twelve-page letter hand-written in English from 'my friend' Gonella to Chagas and mailed from Hartford, Connecticut on 18 February 1986.

He had also gotten a phone call from Chagas. Chagas told him that he was not depressed but irritated at the overall situation. He had received my letter of 27 March outlined above. Chagas asked if Canuto would go to Turin to try to explain the situation to Ballestrero. Chagas would have to find the money to send him (the point is made from time to time that the Pontifical Academy of Sciences is operating on a shoestring). Canuto described Chagas as being 'pissed off' with Gonella.

Canuto asked Chagas if he could read the Gonella letter to me. Chagas said no but he could give me its gist. Canuto described it as the most unfriendly letter he had ever seen—he told Chagas it was a punch below the belt—it should never have been written.

In the letter, Gonella claimed that my activities related to dating the Turin Shroud were motivated to secure my funding from the National Science Foundation. At that point I interrupted with "my God that's ridiculous". Gonella said that I was running the shroud dating enterprise without authority from the other laboratories or the British Museum. Harbottle was annoyed at me. The sample sizes that had been requested were too large and no distinction had been made between the amount required for the counter labs and the accelerator labs. In the discussions in New York, I had made a poor estimate of the areal density of the cloth.

I told Canuto that I had made the best estimate that I could from the information I had on the shroud. If all the idiotic investigations that have been carried out have not even produced an accurate number for its areal density, then I have no high hopes for the results of further tests that Gonella might approve.

Gonella wrote that the Trondheim Protocol was poorly thought out. He had explained carefully to me that there was a very important relationship between the tests STURP wanted to carry out and the actual taking of samples from the shroud.

That was the first I had heard of such a connection. No one had ever explained to me why the sample removal from the shroud had anything to do with the battery of tests STURP wanted to pursue. In fact, of course, there was no such connection.

He wrote that there was fierce competition between National Science Foundation funded laboratories and that nuclear physics was a dying field.

I told Canuto that Gonella sounded very much like my colleagues in particle physics in the Department of Physics and Astronomy here at Rochester.

Gonella claimed that Harbottle had said that I was famous for grabbing the ball and running with it. Canuto injected here he did not know I played football. Gonella stated that somehow I missed the point, that it was not up to me to decide which work was to be done on the shroud and by whom.

Canuto said he wished he could send me a copy of the letter.

Continuing his paraphrase of the letter, he said Gonella noted that laboratories needed lots of money to operate and that Arizona was not running at that time.

I replied that this happened to all laboratories from time to time. We were down right now but soon would be operating again and Arizona was probably running now.

Gonella claimed that dating the shroud would loom high in the eyes of Congress and, in view of the funding problem, it meant the survival of the laboratories. Brookhaven was outside the university competition for funding and Harbottle had told Gonella that Rochester and Arizona were engaged in a struggle for survival and the NSF was sharply reducing funds for nuclear physics laboratories. These accelerators were very expensive machines that yielded little results.

I noted that some of my colleagues in the tandem laboratory at Rochester looked upon the whole thing with some suspicion because they did not think it was going to do great things for the lab. We were funded to do nuclear physics, and the shroud was way off the main stream.

I told Canuto that I regarded Gonella as a trouble maker of the first order and he agreed.

I pointed out that the Trondheim Protocol had been very thoroughly reviewed by all the parties and approved by them.

In his letter, Gonella also disputed the required sample size of 6 square centimetres that I had quoted in the protocol. That amount was based on the best estimate I had at the time of the shroud's areal density. He claimed that the accelerator laboratories needed less than the counter laboratories. This and my incorrect areal density figure would change the amount of cloth needed. I said all such questions could be settled at the Turin workshop. At that workshop, the final amount of cloth required for the seven laboratories that finally became involved was 12 square centimetres, so I had been off by a factor of two. So what!

From what Canuto told about the Gonella letter (I still have not seen it) it was completely malicious and untrue. Nothing like this had ever happened to me before in my scientific career. What were Gonella's motives in writing such a letter? Clearly he was doing his best to think of any reason he could to delay the carbon dating of the shroud—perhaps not so much because he was opposed to carbon dating, but rather because he wanted the dating closely linked to the STURP organization. Perhaps he felt his control over the destiny of the shroud was being threatened by my actions and he wanted to discredit me in the eyes of Professor Chagas. Whatever the reason, he acted in a most unprofessional manner.

Around this time, Monsignor Celli, the Vatican ambassador to the UN, contacted Chagas and suggested that the case for not having a meeting at the Turin Polytechnic was more general than the fact that it was Gonella's home institute. It made no sense for the Academy to hold meetings in a small room in the Polytechnic; that would not be elegant nor would it be traditional. It must be held at the Vatican. Anti-STURP arguments for not having it at the Polytechnic were really irrelevant. Chagas said that he would both write to and phone Ballestrero insisting that the meeting be held in the Vatican. Canuto suggested that I should contact the various laboratories with respect to the question of the meeting being held at the Polytechnic. Would they be willing to meet there? If they were not willing to do so, was there unanimity?

On 2 April 1986, Uznansky phoned me from the National Science Foundation. He said that the reviews were in on my proposal for funding the three-member US delegation to the Turin workshop. There had been high praise for the competence of all the people and the laboratories involved, but there were other questions which caused him problems with respect to the reviews. He had requested clarification from the present reviewers. He said he should know in a couple of weeks and would call me immediately. I told him I would prepare a document that amplified the scientific merits of dating the shroud.

I prepared an addendum to my NSF proposal titled 'Scientific Justification for Dating the Turin Shroud'. I stated that such a dating could only be carried out using very small samples. This was made possible by the development of accelerator mass spectrometry and of decay counting using small proportional counters. Dating the shroud posed the most stringent challenges to these new small-sample dating techniques. No one would argue that such a dating had any substantial scientific significance. It did, however, provide an important test of a new scientific technique. It was an example of how science can contribute answers to questions of wide public interest. One of the missions of the National Science Foundation, after all, was to popularize science.

I wrote Canuto stating I was very disturbed that Gonella, in his capacity as "the trusted representative of Cardinal Ballestrero for the matter of the C-14 test of the cloth of the Shroud of Turin under the patronage of the Pontifical Academy of Sciences" would write a letter to the president of that Academy accusing me of unprofessional behaviour. I said my reaction to such a baseless charge was mitigated by the fact that Gonella had demonstrated, by writing such a letter, that his own professionalism was more questionable than he considers mine to be. I also found it surprising that Chagas would give Canuto a copy of Gonella's letter and refuse to let him read it to me verbatim. I noted that the Shroud of Turin was the greatest thing that has ever happened to Gonella but it was not to me. My impulse was to say to hell with the whole thing except that some pretty high-class people in high-class institutions were now involved. A fevered imagination might conclude that the people connected with some or all of the institutions were motivated by lust, venality, cupidity, self-aggrandizement or other unprofessional concerns, but few rational people would take such a charge very seriously.

I admitted to acting as a spokesman for the six laboratories and the British Museum. It was not a self-appointed position but stemmed from a suggestion they had made that I prepare a protocol that might lead to dating the shroud. What had happened since followed rather naturally. I did not have an official document I could produce for Gonella that authorized me to do anything on behalf of the institutions involved. They had been kept well informed of all developments and so far had not expressed any uneasiness with what I had been doing.

I confessed to having thinly veiled negative feelings about STURP and suspected Gonella was serving STURP's interests as much or more than he was the cardinal's. I had made it clear that I would have preferred the carbon-14 dating to be completely dissociated from STURP. I had also made it clear that if Professor Chagas wanted them involved, they would be involved. As far as I was concerned, Professor Chagas was in charge and his will should prevail. I concluded by stating I would continue to be involved until Gonella's heel nipping became too annoying.

On 3 April 1986, I received a most remarkable document. It was dated 17 March 1986 and was from Robert H Dinegar on the letterhead 'The Shroud of Turin Research Project, Inc.' In it he stated that it appeared STURP would be successful in obtaining samples of the shroud for radiocarbon dating. He said in February Luigi Gonella had informed STURP that its carbon-14 shroud dating proposal had been accepted by the Vatican and the Pontifical Academy of Sciences. As a result, STURP would be allowed to remove samples of the cloth sometime during the upcoming summer for distribution to all participants. Details concerning how many samples would be allowed and from where on the shroud they would be taken as well as the cleaning procedure to be used needed to be settled. As a result, he revealed there would be a meeting of STURP and all other participants in Turin, 9–10 June 1986. More details of this meeting would follow. It was signed by Dinegar.

This letter had been sent to a 'C-14 laboratory distribution list' that named Dinegar as the coordinating investigator, Dr Garman Harbottle as the associate coordinating investigator, and the following 'analytical investigators and consultants': Harbottle, Oeschger, Damon, Tite, Stuiver, Otlet, Hall, Gove, Woelfli and Donahue. As far as I was concerned the most incredible part of the document was a little red arrow that Dinegar put at my name, then in his own handwriting on an attached post-it note he wrote: "A 'slip' now could have a disastrous consequence. RHD". This made my mind reel. Did Dinegar actually believe that he had arranged for this workshop to take place in June, that the workshop was under the control of STURP and was a result of the STURP proposal? I began to wonder whether it was Dinegar or me who was becoming delusional.

The Dinegar letter so outraged me I decided that I would have to consult with Professor Chagas. I first called Canuto, read him Dinegar's letter, and told him I was trying to phone Chagas. His reaction to the letter was much the same as mine. I said I would call him back after I had talked to Chagas. I then called Chagas. He said he had not yet answered my letter because he was still having discussions with Turin. He was expecting to resolve the issues by the middle of next week. I said I was calling him about a letter I had received from Dinegar dated 17 March and I asked him if I might read it to him because it was short. I did so and as I read the phrase, "the meeting of STURP in Turin 9–10 June", Chagas said "but this is nice". I finished reading it and Chagas said "I think this is good, what is your reaction to it?" I said it sounded to me as if STURP were organizing the meeting and not the Pontifical Academy.

Chagas, ignoring this, said he had just composed a long letter to Ballestrero. In the letter Chagas stated that: (1) the meeting would be held in Turin, (2) he would chair the meeting and Ballestrero would be the honourary chairman, (3) he could not have a meeting anywhere other than in the bishop's palace. The Pontifical Academy of Sciences traditionally did not hold meetings outside Rome: he was making an exception for Ballestrero in having a meeting in Turin but could have it only in his palace. The letter would be sent tomorrow by telex.

I asked him whether he wasn't a little troubled by the letter from Dinegar—it seemed to say the whole thing was going to be a STURP enterprise. He said his letter to Ballestrero stated clearly that the meeting was a Pontifical Academy meeting. I interjected that the Dinegar letter said it would be a STURP meeting and when the laboratories received it they would be totally confused. I said I had read the letter to Canuto and he was very disturbed about it. Chagas then said, "I am very disturbed about it because it (the workshop) was organized by yourself, and myself and by Canuto." I said that was correct but this sounded as if it had been organized by Gonella and Dinegar. Chagas then said he would speak tomorrow with Gonella.

I said the letter claimed that in February Gonella said STURP's proposal had been accepted by the Vatican and the Pontifical Academy. Chagas said if it were not an academy meeting, he thought he was going to get the academy out of it. I replied that I thought the laboratories would get out of it also unless it was run by the academy. He said he was very thankful for my support.

I then phoned Canuto and reported to him the gist of my call to Chagas. I said that I thought that Chagas did not fully comprehend the import of the Dinegar letter. Canuto said that he was surprised at Chagas' reaction, but probably he would need to read the letter himself. I said I would send a copy to Chagas. Canuto said he would immediately call Chagas.

Immediately after he called Chagas, Canuto phoned me and opened with "Put some humour in this tragedy! This is becoming a nightmare." He reported that he had talked to Chagas. Chagas said that he had understood my phone call. He was going to phone Gonella. Canuto told Chagas that I would send him a copy of Dinegar's letter and he added that the letter had clearly been inspired by Gonella. I said that Gonella must have been a very busy fellow since Shirley and I had met with him in New York: he had met with Ambassador Celli, visited Harbottle in Brookhaven and the STURP chief in Connecticut; written the twelve-page diatribe against me; inspired the Dinegar letter; and God knows what else. Canuto agreed and said he was very unhappy about the Dinegar letter and was a little uncertain about how to handle it.

I called Harbottle on 4 April 1986. He said he had received the Dinegar letter. I told him that I had phoned Chagas and read the letter to him and he was baffled by it. Chagas had composed a letter to Ballestrero stating that although the meeting would be held in Turin—a concession to Ballestrero—it would have to be in the Archbishop's Palace and certainly not at the Polytechnic. Furthermore, it would be chaired by Chagas. Harbottle agreed that made sense but then said it was a mistake to have gotten the Pontifical Academy involved in the first place! I was amazed by this remark.

He went on to say they were merely consultative to the pope and had no connection with Turin. Despite the fact that the shroud belonged to the Vatican, a few years ago when it was suggested that the shroud be moved to Rome all hell broke loose in Turin. Although the pope owned the shroud, it was the cardinal archbishop who must be dealt with. I remarked that it seemed to me that the Pontifical Academy could deal with the cardinal archbishop a lot better than we could. He replied the academy was not of the church—it was merely an honourific advisory body to the pope on questions of science. I replied that this was not true—their headquarters were in the Vatican—they were officially part of the Vatican. That surprised Harbottle but he continued to insist he foresaw nothing but trouble in having them involved.

He said that there had never been any question in his mind that this meeting in Turin, that would be basically a protocol and discussion meeting, was not being arranged by STURP but rather by Ballestrero, Gonella and the academy. He suddenly realized that he had not read the letter carefully enough and he read back the last line: "As a result there will be a meeting of STURP and all other participants on the 9th and 10th of June. More details of that meeting will follow." But he went on to argue that it was just Dinegar's statement, not that of the archbishop or Gonella or the academy.

I noted that Gonella had met with Harbottle, then gone on to Connecticut to meet the STURP boss D'Muhala. He still had had time to write a scurrilous twelve-page letter about me to Chagas. I noted that, in this letter, he quoted Harbottle as saying that there was serious infighting among US nuclear laboratories, nuclear physics was a dying field, we were fighting for our lives. My unseemly grabbing for publicity and leadership in the quest to date the shroud was motivated by a desire to impress the NSF to continue funding for my laboratory.

Harbottle seemed taken aback and finally said, Gonella "screwed up on that one". He said that he had told Gonella that two years ago Teddy Hall was about to have his Van de Graaff laboratory phased out by the British National Research Council. Hall was trying to capture the whole shroud enterprise for Oxford to save his neck. I pointed out to Harbottle that the Van de Graaff laboratory at Oxford was run by Ken Allen. Hall had nothing to do with it. Harbottle said that in any case Hall was the power behind the throne—he was the guy with the clout. He was a member of the Royal Society and the House of Lords and all the rest of it (Hall was neither). I injected "Garm you are a bit confused".

It finally turned out he was not referring to the Nuclear Physics Laboratory at Oxford but to the Laboratory for Archaeometry and the History of Art located next door and run by Hall. Harbottle said Hall's laboratory had bought one of the General Ionex machines and "they never could get the bloody thing to work for years and years. The damn thing didn't produce anything and the British Government was getting fed up with it. They kept pouring more and more money into it." As a result, Harbottle continued, Teddy tried to capture the Shroud of Turin for the Oxford archaeometry lab. I replied, whether that was so or not, Hall was not acting that way now and Harbottle agreed. He denied that he ever said that I was using the shroud dating as a way to get funding. We agreed that Gonella was causing a lot of trouble. He was not a major scientific figure but he had one big thing going for him. He had the archbishop's ear.

I then suggested that it might be a good thing if Harbottle resigned from STURP. He said that he was merely trying to keep his feet in both camps—his only connection with them related to the carbon dating because he thought they needed a good strong dose of scientific rigour in that area. He reminded me that it was he who organized the meeting in Bradford, England that precipitated the British Museum laboratory intercomparison that he modestly characterized as "a helluva good idea". (Harbottle had told me in April 1985 that this was Otlet's idea.) I pointed out that, if he had, he had not done it as a member of STURP.

I then suggested that he should call his friend Dinegar and try to get him straightened out. He replied, "As a matter of fact, now that you have called it to my attention, I will certainly straighten him out on this point, that he should not be giving it out that this meeting in Turin is a STURP meeting, to the contrary." I said I hoped he would and added, "Try to picture what these other labs are going to think when they see this particular document. They will wonder if we are all going mad."

He said he would be seeing Dinegar in a couple of weeks at a STURP meeting that was scheduled to be held in Dayton (it was the first time I had heard of this meeting but, after all, I was not privy to their secret conclaves) and he would try to present to the STURP membership, from the point of view of the laboratories, what he hoped was going to happen in Turin. I urged him to make clear to them that it really was a Pontifical Academy show. I admitted that I would like Chagas to show a little more leadership, but that he was in a difficult position—cardinals are very powerful people. Harbottle agreed.

I repeated that if the Pontifical Academy was not going to be the sponsor of the whole shroud dating operation, that would end my involvement. He said he would hate to see me withdraw and if having the meeting under the auspices of the academy would keep me in, he would certainly plump for that. He certainly did not want it under STURP auspices nor did he want STURP to be the ones to take the samples. That should be done by the lady from Switzerland with someone from at least one of the carbon dating laboratories present.

After this conversation with Harbottle, that must have lasted for at least an hour, Canuto called. He said he had received a call that day from Chagas. Chagas had phoned Gonella with respect to Dinegar's 17 March letter and Gonella said that he knew nothing about it. Canuto told Chagas that that must not be true and Chagas agreed. Chagas told Gonella that the meeting would be in Turin, but would be in an archdiocese' location. It would be run by the academy. Gonella agreed to that. He told Gonella that if Gonella wished to suggest names of STURP people to be present, he should send them to Chagas and he would decide whom to invite. Chagas also told Canuto that samples would be taken on the last day of the meeting.

On 4 April 1986 I wrote to Professor Chagas enclosing a copy of the Dinegar letter for his records. I said that I found Gonella's professed ignorance of the letter incredible. He was kept informed of everything that STURP did. I said I thought the Dinegar letter would cause great confusion in the minds of the people associated with the six laboratories and the British Museum. I urged him again to send all of us official invitations on behalf of the Pontifical Academy of Sciences to attend and participate in the workshop as soon as possible. This would help mitigate the mischief done by Dinegar. I enclosed a list of the names of the people I thought should be invited, including the textile expert Madame Flury-Lemberg. The list of names included Sheridan Bowman, Shirley Brignall, Vittorio Canuto, Paul Damon, Doug Donahue, Teddy Hall, Garm Harbottle, Robert Hedges, Robert Otlet, Mike Tite, Willy Woelfli and me.

I received a short letter from Teddy Hall dated 8 April. He had received various communications from me including my letter to Chagas of 27 March (this was the one in which I said that I was getting the strong impression that the authorities in Turin thought that we all had an overpowering desire to date the shroud and that they could therefore call the shots). He said he agreed with me on this. He confessed that he also had grave reservations about STURP. He wondered whether the fact that Harbottle had gotten involved with them should worry us? He also wondered, having received my letter, whether anything will happen in June. He still hoped it would but predicted problems.

On 15 April 1986 I phoned Uznansky at the NSF. He said he was making progress on our proposal for funding travel to Turin. He had circulated my addendum to the reviewers. He still hoped for the best. He may have to consult Harvey Willard, director of the NSF's nuclear physics programme. There was no question about the competence of the people involved. The real question was the validity of the project. He said he would send me the reviews when NSF action was taken on the proposal, either pro or con.

Harbottle called in connection with the request he had made to various carbon-14 dating laboratories to share with him their experience on the use of charred cloth in establishing dates.

This was triggered by the concern of Meacham, the Hong Kong archaeologist, over the use of charred cloth. He said that a definitive test was going to be carried out by Austin Long at Arizona. He said he had heard that Ballestrero was very angry that the Pontifical Academy was taking things away from him and that the cardinal was appealing to the Vatican secretary of state and to the pope. How Harbottle learned about matters of this kind was a mystery to me.

I contacted Chagas in Brazil. He read me the letter of invitation he was sending to all the participants. Chagas said he had received a cable from Gonella asking him to invite Dinegar, Meacham, and Jacques Evin, a specialist in carbon dating from Lyon, France. I asked if they were the only ones Gonella suggested and Chagas said no, there were also Alan Adler, from Connecticut State University in Danbury, and J Druzik of the J P Getty Museum. He did not think he was going to invite either of them. He said that he would add to the letter of invitation that the meeting must be held in the Archbishop's Palace or in some building in the archdiocese. The invitation letter would be sent to all of the people on my list and, in addition, to Dinegar, Meacham and Evin. The latter three did attend. Adler and Druzik eventually were invited and Adler came but Druzik did not. A person named Steven Lukasik came instead.

Chagas said he was furious about Dinegar's letter. I replied I was not surprised; I did not understand that letter at all. Then Chagas said, "I think in my letter of invitation, everything is put in its place. Don't you think so?" I said yes, I did think so. Towards the end of the conversation I said I did not understand why he was inviting Evin and Meacham. Chagas replied that it was a question of political arrangements with Turin. It was a very difficult situation for him.

The letter of invitation from Chagas to the various people I have just named arrived a few days later. It was dated Rio de Janeiro 15 April 1986.

"Dear Professor Gove, This is to invite you to attend a meeting organized by the Pontifical Academy of Sciences to be held at the Archbishop's Palace, in Turin, from 9–11 of June next. The meeting will be held under the high patronage of His Most Reverend Eminence Don Anastasio Cardinal Ballestrero, Archbishop of Turin, who has given so much of his time and interest to this important subject.

"The workshop will be chaired by myself. Its purpose is the radiocarbon dating of the Shroud. This will need the establishment of a final protocol. This protocol will include all the details regarding the obtainment of the minimum-sized samples, its distribution by a responsible institution to the specialized laboratories represented in the workshop, the analysis of the methods used for dating, as well as the publication of the results.

"The workshop will also decide, with the approval of H.R.E. Cardinal Ballestrero, the responsibility for the follow up of the agreement reached at the meeting.

"I hope that you may obtain travel funding from your institution or some other source in your country, e.g. the equivalent of the National Science Foundation in the USA. If not, the Pontifical Academy of Sciences will cover your travel expenses as well as your sojourn in Turin.

"I pray you to keep this invitation confidential as this is needed for the complete success of the enterprise.

"Hoping to have your acceptance as soon as possible, I am yours very truly, Carlos Chagas."

On 13 April 1986, an article by Peter Jennings appeared in the US Catholic newspaper *Our Sunday Visitor*. I did not receive a copy of it until mid-May after several of the events described below occurred. Jennings was the writer/reporter I had been warned by David Sox in Turin in 1978 to be very careful about talking to. He had called Tite at the British Museum in late February, 1986, to find out what was happening on the shroud dating front and Tite told him that negotiations were proceeding. His article was headlined 'Vatican may allow carbon tests on Shroud of Turin'. Jennings described the interlaboratory tests coordinated by the British Museum and carried out by six radiocarbon laboratories. He said these tests might persuade the Vatican to allow a radiocarbon measurement on the shroud to proceed. He said the results of the British Museum tests would be published in the journal *Radiocarbon* and that they demonstrated that reasonable results can be obtained when several laboratories undertake blind measurements. Richard Burleigh, who coordinated the tests, told Jennings that the shroud age could be determined to an accuracy of plus or minus 50 years but would only show the date the linen flax died and not how the image got on the cloth.

I thought the article was largely benign. However, there was one quote in it, arising from a telephone interview that Jennings held with Professor Carlos Chagas. It said that the President of the Pontifical Academy of Sciences, Professor Carlos Chagas, told *Our Sunday Visitor* from his home in Rio de Janeiro in Brazil, that the Academy was having a meeting in June to discuss the possibility of allowing threads to be taken from the shroud for carbon testing. Professor Chagas emphasized that it is only a preliminary meeting, but, if a decision was taken to allow the tests, then he would be happy for the British Museum to act as coordinator and that the threads from the shroud would be given to the six laboratories that had taken part in the interlab comparison tests. It went on to say that Professor Chagas revealed that a textile expert would be present at the meeting in either Rome or Turin. He disclosed that, personally, he was against a test.

The parts of the article that were most troublesome were that the meeting would be held in June (Chagas had requested we keep this information confidential) and that Chagas was not in favour of having the shroud dated. Professor Chagas had told Shirley Brignall, Vittorio Canuto and me, in the first meeting we had with him in New York, that he personally would prefer that no carbon dating test be made of the shroud. However, there was considerable interest in establishing the date of the shroud both on the part of the general public and of organized groups such as the British Turin Shroud Society and of many scientists. He therefore had concluded that the carbon dating just had to be carried out but he wanted to ensure that it be done in the best possible way. I would be very surprised if he did not make that same comment to Peter Jennings. If he did, Jennings might well have found it useful not to include it in his report.

On 24 April I phoned Uznansky inquiring about the status of the NSF proposal for travel funds. He indicated that he was still waiting for the comments from one final referee. He had sent this proposal to seven referees, one of whom gave a very poor review—there was no scientific merit to the proposal. He said that this reviewer argued that questions of authenticity of art should be privately funded and that it was a matter of complete indifference scientifically how old the shroud was. He would get back to me as soon as he had more information. I was astonished that a proposal for some six thousand dollars would have to be reviewed by seven people. A proposal to fund a space shot to land monkeys on Mars would need half that number.

On 30 April I wrote a letter to all of the laboratories saying that by now they should have received an invitation from Chagas to attend the June workshop. I attached a list of the names of the people I had submitted to Chagas, all of whom were invited. I noted that there were other people who would be invited as well, and these included Luigi Gonella, Robert Dinegar, Jacques Evin from a radiocarbon laboratory in Lyon, France, Bill Meacham from the Hong Kong Museum of History and Madame Flury-Lemberg. I added that it was Gonella who suggested the names of Dinegar, Evin and Meacham, but that he had also suggested a couple of other people, who were not accepted by Professor Chagas. I noted that I was pleased with the outcome so far, particularly the fact that it would be under the auspices of the Pontifical Academy and would be held in the Archbishop's Palace in Turin and not, as originally suggested, by Professor Gonella at the Turin Polytechnic. I also noted that it would not be a workshop run by STURP as suggested in Dr Dinegar's baffling letter dated 17 March. I concluded that it would be useful if we could agree on an agenda prior to the workshop and the enclosed agenda I had prepared was an outline of what might be discussed.

It was on 10 May 1986 that the first inkling that something might have gone awry occurred. Harbottle phoned to say that Gonella had called him. Chagas had given a phone interview from Brazil to Peter Jennings. Chagas told Jennings that a preliminary meeting would be held in Turin in June concerning carbon dating the shroud. Harbottle was terribly upset. He asked how Chagas could have done this after having enjoined all of us to strict secrecy in his letter to us inviting us to the Turin workshop? He said that he feared that right wing terrorists would be lying in wait for us in Turin. I do not consider myself particularly fearless but this struck me as ludicrous. I wondered if Harbottle was beginning to lose his grip.

I contacted Canuto after the conversation with Harbottle and told him about the Jennings interview. Canuto then called Chagas. This interview with Jennings occurred about a month ago. Jennings asked Chagas if he were organizing a meeting and Chagas said yes, it would be held either in Rome or Turin in June. He didn't say any more. He said he didn't want to lie to Jennings.

On 13 May I received a letter from Uznansky saying that he was pleased to confirm that the National Science Foundation and the Pontifical Academy of Sciences had approved the joint workshop on radiocarbon dating of the Shroud of Turin that I had organized with Professor Carlos Chagas. Included in this letter were the reviews from the referees. There were a total of six referee comments—the seventh never responded. Such reviews are sent to the principal investigator verbatim but the referees are anonymous. The overall ratings requested by the NSF range from excellent to poor.

The first one rated the proposal as fair—not a high rating. The referee said he had the highest regard for the AMS programme at Rochester and, normally, would rate a proposal submitted by Professor Gove highly. However, he thought, in this case, there was little scientific merit in the carbon-14 dating of the Shroud of Turin, and that it would be inappropriate for the NSF to fund the workshop.

The second referee rated the proposal as somewhere between excellent and very good. He said that the discussion of procedures was most important, so he strongly supported the workshop.

The next referee said that he was unqualified to make any judgment despite the fact that he had the highest regard for the scientific competence of Gove, Donahue and Harbottle. He also knew Hedges and Woelfli and regarded them highly as well.

The next one rated the proposal as excellent. He said that Gove was going about the dating of the shroud in a very exact and non-controversial manner, as necessary for this relic. Anything short of this and the whole operation might become suspect. He was asking for a bare-bones budget and the request should be supported in its entirety. There was no question of the competence of the people chosen for the workshop.

The next referee rated it poor. This was the one that was troublesome. The reason given was that the question being addressed was one of authenticity, not science. There was nothing weighing in the scientific balance that depended on the age of the shroud. It was a religious relic and nothing scientific would be as learned by dating it. It was analogous to dating a supposed Rubens. Art authentication should be supported by private funds. I was reasonably certain that the review came from my friend of long standing, Roy Middleton, of the University of Pennsylvania. It was a legitimate reason for giving the proposal a negative rating but it displayed a certain naivety—it had a typical ivory tower quality to it.

Finally, the sixth reviewer rated it excellent. He said the workshop was without question a very important event. It promised to open the way for a long-awaited age determination of the shroud. The result would have religious and historical significance and would almost certainly be challenged by many if the procedures were not prepared with the utmost care.

Uznansky consulted Harvey Willard, who was in charge of the nuclear physics programme at the National Science Foundation, and Willard concluded that the one poor rating was made on an inexact argument. The Turin Shroud was not like any other art object whose authenticity was dubious but rather it had both a religious and historic importance. It went far beyond a simple authenticity test of an art object and therefore would be quite a proper enterprise for the National Science Foundation to fund.

If anyone doubts the care with which the NSF spends the taxpayers' money, this example should allay their concerns.

Chapter 5

The Postponement Frenzy

A most flabbergasting event in the shroud saga occurred on 16 May 1986. I received a cable signed "Rovasenda, Director of the Pontifical Academy of Sciences" that read "Regret to advise that owing (to) organizational problems meeting June 9 to 11 postponed. Stop. Letter from President Chagas follows."

It was both unexpected and unacceptable. I thought it outrageous that less than one month before the proposed workshop was to take place we received a peremptory notice that it was postponed. I suppose I should have had some inkling of this possibility from some previous events, in particular, the fact that Luigi Gonella, in his phone call to Harbottle of 10 May, professed to be very concerned about what I thought was a relatively innocuous interview that Professor Chagas had given to Peter Jennings. In any case, it triggered a frenzy of activity on my part that lasted from 16 May until 30 May with something like forty or more phone calls being placed to various people. I decided that, rather than supinely letting this pass, I would try everything I possibly could to discover why the meeting had been postponed and to get the order for postponement rescinded. Despite all my efforts, I failed and the meeting was indeed postponed. Contrary to my predictions that this meant permanent cancellation, it was eventually rescheduled for 29 and 30 September and 1 October 1986.

On 16 May 1986 I flew to Boston to attend a retirement symposium for a colleague, Harald Enge, at MIT. I called Vittorio Canuto from the Space Science Center at MIT where the meeting for Enge was being held and described the contents of this cable. He speculated that it was a result of the Chagas interview with Jennings. Canuto said Chagas was visiting his daughter in Paris.

119

I then phoned Luigi Gonella at his home in Turin. His wife answered and said he would be home about 8 pm Turin time.

I talked to Canuto again. He told me he was trying to phone everyone he could think of but without much luck. He said that Ambassador Celli had called Monsignor Rovasenda, the Chancellor of the Pontifical Academy in the Vatican. He was told the meeting had been postponed on the orders of the Secretary of State, Cardinal Casaroli, because of a letter he had received questioning the planning procedures for the workshop. Canuto said that Chagas would be going to Rome from Paris to read the letter. He had managed to contact Donahue and Donahue reported that he had read the report by Peter Jennings in *Our Sunday Visitor* and he was troubled by it. However, at this stage, it was not clear that it was this news report that had caused the postponement.

Later that afternoon of the meeting at MIT, I met with Professor Victor Weisskopf, who was a member of the Pontifical Academy of Sciences. I had known Viki for some forty years. He had been one of my professors when I was studying for my PhD at MIT in the late 1940s. He did not seem particularly surprised that the workshop had been postponed. His guess was that the authorities at Turin would go to almost any length to put delays in the way of establishing a radiocarbon date for the shroud. He gave me the phone number of Chagas' daughter in Paris.

I made several attempts to phone Chagas, and finally on 17 May I called his daughter's house and he was there. He said the meeting had to be postponed, the trouble was in Turin and he did not know why. He would be in Rome on the following Sunday, 25 May, and would contact me then.

I reported this conversation to Canuto and he said he would call Chagas. He did and phoned me back. Chagas said that he received the order to postpone the meeting from Cardinal Somolo, the Vatican undersecretary of state. Canuto said that he raised hell with Chagas by saying, "There you are lolling around in Paris while a revolution is breaking out here". He emphasized to Chagas that we couldn't wait a full week until he got to Rome.

That afternoon, 17 May, I managed to contact Gonella at his home in Turin. I asked him what had caused the postponement—had the request come from Turin? He sounded rather flustered. He said that the first actions were by the Vatican. He admitted that Turin had played some role although he did not specify what it was. The reasons however were, first, the article by Peter

Jennings in the Catholic weekly, *Our Sunday Visitor*. He said it quoted Chagas on the time, place and purpose of the workshop. He claimed that this report greatly disturbed the cardinal. The second reason was that Chagas had not invited all the people that the cardinal had suggested (he meant that he had suggested), for example, Adler and the J P Getty man, Druzik. Thirdly, it was the cardinal's meeting, not a meeting of the Pontifical Academy of Sciences, but Chagas had steadfastly refused to concede this. He suggested that Chagas was trying to play too dominant a role.

I said he must realize that this postponement was a serious matter. There were major institutions involved, for example the British Museum, the National Science Foundation, Oxford, Rochester, Arizona, Zurich, and two national laboratories, one in the US and one in the UK. All of these would have their worst suspicions confirmed—that the Catholic Church was reluctant to have the shroud dated. He must persuade the cardinal to reverse the postponement order. He said he would pass my comments along to the cardinal. He continued to lay all the blame on Chagas and would not admit that Turin precipitated the postponement. He kept insisting that I should get information on all of this from Chagas. I said that I had talked to Chagas within hours of this phone call and that he knew nothing about the details. Gonella denied that this was in any way the end of shroud dating.

On Sunday, 18 May I described my conversation with Gonella to Canuto. He had talked to Ambassador Celli. Celli described Peter Jennings in a rather unflattering way, and said that such a trivial article should not end the plans for dating the shroud. On the other hand, it would likely cause the scientists involved to contact the press and raise hell. In Celli's view the Church was not at all afraid of testing the shroud, but that would certainly be the way this postponement would be interpreted.

I then decided that a cable, signed by as many people as were involved in this workshop, should be sent to Ballestrero in Turin and Chagas and Casaroli at the Vatican. I mentioned the idea to Canuto and he said that I should first consult Chagas.

Chagas called me from Paris shortly afterwards. I described my phone call to Gonella who told me that the postponement had been initiated by the Vatican, and that Turin had played some role—he didn't say what. I gave the reasons for Turin being unhappy about the workshop. I said I had prepared a cable to be sent protesting the postponement. He said that he was going to Rome next week. Ballestrero would be there and Chagas was confident that the matter would be resolved. He agreed that we should send a protest cable.

At this point I needed to read the contents of the proposed cable, that I had now composed, to as many people as possible in order to get the largest number of signatures. I phoned Donahue and read it to him.

"The last minute peremptory postponement of the June 9–11, 1986 workshop in Turin to discuss details and procedures concerning radiocarbon measurements of the Shroud of Turin will have consequences those who ordered it may not have realized. Two of these are: 1. A number of the key institutions involved will likely withdraw from the project (the institutions were listed). The organizational effort in assembling this collection of undeniably meritorious institutions was substantial. Several of them had to obtain permission from their top management to be involved in the project. Their participation has now been placed in extreme jeopardy. 2. The postponement of this carefully organized workshop, the planning for which has been underway for almost a year, will be tantamount to termination of the project, at least in the near term. There are no credible reasons that could possibly be advanced at this late stage for postponing the workshop. No reasons credible or otherwise have yet been given. Whatever the Vatican or the Archdiocese of Turin find troublesome about details of the workshop should be readily resolved without resorting to a last minute draconian postponement. The undersigned representatives of some of the institutions invited to participate in the workshop urge that the postponement order be rescinded immediately and that the workshop proceed. We reiterate our view that the postponement will have extremely deleterious consequences of which its proponents are apparently unaware or, what is more incredible, about which they are unconcerned."

Donahue said that he would be quite willing to sign it.

The next morning, Monday 19 May 1986, I phoned Tite at the British Museum but he was attending an archaeometry conference in Athens. I tried to call Woelfli in Zurich but there was no answer at his institute—it was a national holiday in Switzerland.

The next person I contacted was Harbottle. He was quite happy to have his name attached to the cable. He expressed concern about sending it by public telegraph. He said that it should either be sent by code or by diplomatic pouch. Otherwise it should be reworded in such a fashion as to omit the proposed dates of the workshop. Harbottle was being extremely cautious about maintaining secrecy concerning the time, place and purpose of the workshop—unnecessarily so in my opinion—so I ignored the advice.

I next got through to Teddy Hall at Oxford. He said that he would be quite willing to sign for Oxford. Furthermore, he said he would be willing to phone the Director of the British Museum, Sir David Wilson, to see if Wilson would be willing to sign on behalf of the British Museum. He phoned me back a little later to say that I could add Wilson's name as a signator. He was the sort of person who didn't mind sticking his neck out. I should not include Tite's name because he was not available to give his consent. Hall said that he didn't think the cable would make any difference. He said that they were very unlikely to rescind the postponement order. He was right on the money about that.

On 19 May the cable was sent to Ballestrero, Casaroli and Chagas via Western Union. It was signed by Donahue, Harbottle, Wilson, Hall and me. I then dictated the cable over the phone to Canuto for him to pass on to Ambassador Celli.

Later that afternoon, I again talked to Canuto. He had read the contents of the cable to Ambassador Celli. Celli said its wording was far too strong for him to have sent to his boss, Cardinal Casaroli but, for those who signed it, it was by far the best way to proceed. I told Canuto that I was beginning to think that the postponement was arranged so that the meeting could actually be held in connection with the tests that STURP was proposing. Canuto recalled that he had suggested this possibility when he talked to Harbottle.

Canuto called me again the same day. He had talked to Mrs Chagas. She told him that Ballestrero was going to meet with Rovasenda tomorrow, 21 May. Rovasenda had the cable that had been sent to Chagas at the Pontifical Academy of Sciences headquarters in the Vatican. Chagas said that his plans were to go to Rome on Sunday 25 May but depending on the conversation between Ballestrero and Rovasenda he might go earlier.

Around 3 pm on 20 May Canuto called again and said that he had spoken with Chagas. Ballestrero was now in Rome. Chagas might go to Rome before Sunday, but who knows? Canuto said that Celli was going to contact Rovasenda and Canuto also wants to talk to him and will call me back.

Then about an hour later, he got back to me to say there was very bad news. Rovasenda said the order for the postponement came directly from the pope. Rovasenda was completely on our side and realized all the ramifications that might result from this postponement. The Vatican might look very silly for having postponed the workshop on very flimsy grounds. However, since the order was from the pope that was it as far as he was concerned.

That same day, I received the following cable from Professor Chagas: "I am actually negotiating. Ask you to be patient. Will cable you final decision next Tuesday May 27 or Wednesday, best regards Carlos Chagas." This cable suggested that there might be some consequence from our 19 May cable and the decision to postpone the workshop might actually be reversed. It also sounded a bit testy. Clearly he was trying to show that he was not just 'lolling around' but was taking action.

On 21 May I phoned the J P Getty Conservation Institute in Venice, California, to see if I could find out who Druzik was. I was told he was James Druzik of the Scientific Department, who had some expertise on testing and cleaning of art items.

It was around this time that I got the idea of trying to enlist the aid of various ambassadors, particularly ambassadors to the Vatican, both those from the US and possibly from the UK. I phoned Teddy Hall, to see if there were some way that he could make contact with Great Britain's ambassador to the Vatican. I was told Hall had left for Athens and would not be back until the end of May and would be on holiday after that.

I contacted Sir David Wilson, Director of the British Museum, and talked to him about this possibility. He said he saw absolutely no hope of going through the UK ambassador to the Vatican. It's the kind of thing that ambassadors really shy away from. He said it might only work if the ambassador were a drinking buddy of the pope. He felt that even going through the US ambassador to the Vatican would be fruitless. He thought the only hope was the cable that we sent and I thanked him very much for his being willing to co-sign it. He thought that the whole affair was preposterous. He suggested that I should work through US–Vatican connections. I said that we had already been working through Monsignor Celli, who was the Vatican's ambassador to the United Nations.

I then tried to contact the United States ambassador to the Vatican, William Wilson. It was Celli who suggested that I might ask Ambassador Wilson to contact Rovasenda, the chancellor of the Pontifical Academy of Sciences. Rovasenda could explain to him what the problem was and Ambassador Wilson might be able to indicate that he thought the postponement should be rescinded.

I attempted to get through to the ambassador directly by calling his embassy in Vatican City but was told by a secretary that he was away and would be back the next day, 22 May. The embassy was just about to close. I phoned the State Department in Washington to get Wilson's home phone and they suggested I call their Foreign Lounge which, in turn, referred me to their Vatican Desk. I was told they could not give me Wilson's home phone number.

On 22 May at 6:50 am I phoned Wilson again at his embassy in the Vatican and was told that he was not in but that he might be contacted after 28 May. They told me that he was *en route* to Venice with his wife and would be staying at the Gritti Palace Hotel. I decided this approach was probably hopeless.

Shirley reminded me that one of our graduate students in the Department of Physics and Astronomy, Michele Migliuolo, was the son of the Italian ambassador to Egypt. I talked to Michele and he said his father was in fact quite a close friend of Cardinal Casaroli. Michele said that he would call his father in Egypt to see if there was any possibility that he might exercise some influence. I was clearly grasping at straws at this point. Michele did call his father and his father said that it was very difficult to do anything about it from Cairo. He felt that the pope must have good reasons for this postponement, particularly since it was given on such remarkably short notice. He said that he would call Casaroli to see if he could get any information from him and then call Michele the next day.

I heard later on that day that Canuto had called Casaroli. Casaroli was very friendly: he apparently knew nothing about this postponement; it was another branch of the Vatican State Department that was dealing with it. I assumed this was Cardinal Somolo, the undersecretary of state.

Around noon on 23 May I again talked to Canuto and he said that he had spoken to Rovasenda at the Vatican that morning. Rovasenda had contacted Ballestrero and reported him as saying that a group of sindonologists in Turin did not want the carbon-14 dating to take place. They were putting pressure on the pope. Ballestrero, however, was not opposed to carbon-14 dating of the shroud. On Monday morning 26 May at 11:30 am, Chagas would meet with the undersecretary of state, Martinus Somolo. Canuto advised us to hold off on changing our tickets to Turin.

Canuto speculated that representatives of the International Centre of Sindonology in Turin had contacted the pope, through Somolo. Rovasenda felt that the main trouble was the Jennings news article—the opposition of this sindonology institute was secondary. That was the impression he got from Ballestrero.

I called Canuto again. He asked me if I had informed Dinegar of the postponement. I said I had not. He said that he would use this as an excuse to call Dinegar and play dumb to find out what he could learn from him about the postponement.

I talked to Canuto again around 6 pm and he told me that he was not able to contact Dinegar. He wondered rather plaintively why should he and I be the only ones who cared anything about this postponement. Chagas, on the other hand, was probably eating good food in Rome apparently unconcerned.

Canuto called me around noon Saturday 24 May. He had received a copy of the Jennings article from Donahue. He said he would give it to Cardinal Baggio, governor of the Vatican State, who was visiting the Holy See Mission to the UN in New York. Baggio was leaving for Rome the next day and would deliver the copy to Chagas. Canuto thought the most damaging statement was Chagas' admission that he personally was not in favour of carbon dating the shroud. Despite this, the article could hardly be used as an excuse for postponing the workshop.

The *New York Times* of Sunday 25 May carried an article headlined 'Vatican Envoy is Coming Home'. It said William A Wilson was an old California friend and advisor of Ronald Reagan who had appointed him presidential envoy and later United States Ambassador to the Vatican. Mr Wilson had resigned from that post the previous week however, amid concern, administration officials said, about unauthorized dealings he had had with Colonel Muamar L Kadaffi, the Libyan leader. No wonder I was unable to contact him. Even if I had, carbon dating the Turin Shroud would not have loomed high on his list of priorities.

On 27th May I called Ambassador Celli. He told me Chagas had met with the undersecretary of state the previous day, but he had no word of the meeting. Canuto called and said that he had not yet heard from Chagas—there was nothing we could do. When and if word came, it would come to Celli and he would contact Canuto's wife at the United Nations where she worked. He said I should try to call Dinegar. I did place a call to Dinegar and I was informed that he was travelling and would be back on 8 June. I presumed he was honeymooning in China with his new bride. I left a message for Ray Rogers to call me.

That evening Canuto called. He had just received a telex from Chagas. I got the same telex the next day. It read: "Holy See and Cardinal Ballestrero have no intention not to date shroud. Adjournment of meeting should only be considered temporarily postponed. Shall contact you in the near future. Best regards Carlos Chagas."

On 28 May I phoned Chagas in the Vatican. He began by asking me if I had received his cable and I said that Canuto had read it to me last night. I wondered whether he could give me some clue as to why the meeting had been postponed. Chagas said he really did not know—there was a group in Turin who raised a fuss, Ballestrero became concerned and so he asked that the meeting be postponed. He said he was expecting to see the secretary of state and the pope in the next days to find out exactly what postponement means. I remarked that the three of us in the USA had already received our airline tickets—what do we do about them? He replied that he would let me know in ten days. I said the tickets would expire before ten days—I was the principal investigator of the NSF grant and had certain responsibilities. Chagas broke in with, "You have as much responsibility as I have I would say. I would ask you to keep quiet for a week. I will phone or cable you again in a week." He said if the meeting were to be held at the Vatican there would be no problem. Then why not just hold it there I asked. He replied, "This Cardinal Ballestrero insists it be held in Turin. What I am fighting for here is to convince him that we should have it here (in the Vatican) because the shroud is not going to be touched." That was certainly new and important information! I asked him whether the International Sindonology Center in Turin had any connection with the Archdiocese? He replied that they were fighting the Archdiocese. Why does the archbishop not ask them to stop, I asked, naively, and he replied that it was very difficult to understand some aspects of Italian politics. I returned to the problem of what I should tell Arizona and Brookhaven. Again he interrupted, "Say I phoned you saying the meeting has been postponed and that I am fighting to make it in Rome". He then said, "My idea would be to postpone it until the end of the year. What would be your reaction to this?" I said that would certainly give people time to re-arrange their plans but it would not be a good time for people in academic institutions. He said he would keep in touch and should have an answer before 15 June. He was now stretching the ten days into over two weeks! At this point I gave up and told him I would inform the NSF of the delay. He said that he would call Uznansky tomorrow. It was a thoroughly unsatisfactory phone call.

I then contacted Celli and told him about my conversation with Chagas. He said there must be some prominent sindonologists who caused the delay. I phoned Uznansky at the NSF and he said that as far as the grant proposal was concerned, it would still be valid, but I should call the travel agent to cancel the tickets.

Around 4 pm that afternoon of 28 May Ray Rogers phoned me from Los Alamos in response to the message I had left. He said that he would not be involved in the upcoming STURP tests in Turin. He had been dropped from STURP because he was not a 'soldier for Christ'. He said that there were two other people who might be involved in the test, Larry Schwalbe and Bob Dinegar. He said that he was very disillusioned about the STURP organization. He said that their publications of the 1978 measurements were essentially worthless. He said that there were two people in the Turin Sindonology Centre who were fanatics. One was a man named Baima Bollone and the other was Don Piero Coero-Borga. He reminded me that Ballestrero had said back in 1978 that when the laboratories agreed on a method for carbon dating, that samples would be taken. The Sindonology Centre was probably now saying we want to use two methods and that means we cannot agree on a single method for dating of the shroud. Thus the tests should not proceed any further. Rogers said that carbon-14 dating was the absolute watershed experiment. The other tests just meant nothing. He left STURP because association with that organization and what it had become would tarnish his reputation. He felt that it was very important that absolute security be applied to shroud samples, that there must be no possibility of substitution. The only arrangement that would satisfy him would be for the senior members of each lab to take the samples back to their labs after witnessing the removal. I was saddened to hear that Rogers had left STURP because he was one of the few scientists in that organization for whom I had real respect.

On 30 May Canuto called. He said he just did not find it plausible that the International Centre for Sindonology in Turin could single-handedly have gotten the workshop postponed. I told him that I would be giving an update on the shroud dating at a workshop to be held at Oxford University in June, and I thought that I would just present a set of slides quoting the cables that had passed back and forth on the occasion of this postponement. He said he hoped that I was kidding.

Shirley and I had rearranged our travel plans so that instead of going to Italy we would be going to Portugal and then to England. There I would be giving a paper at the Royal Society and then attending a workshop on accelerator mass spectrometry at Oxford University.

On 10 June I wrote a letter to Carlos Chagas telling him that I was leaving on 14 June to give a talk in London to the Royal Society on AMS and later, in Oxford at an AMS workshop, a talk on 'Current Proposed Procedures for Dating the Turin Shroud'. While in London, I would be meeting with various people at the British Museum. I gave him a list of the hotels I would be staying at along with the dates and their telex numbers.

By this time, I had been able to obtain a copy of the 13 April 1986 edition of *Our Sunday Visitor* and confirmed that Jenning's article was pretty innocuous. It said that Professor Chagas had revealed that a textile expert would be present at the meeting in either Rome or Turin in June. However, Chagas disclosed that personally he was 'against a test' and the article just left it there.

On Saturday 14 June 1986, Shirley and I left Toronto for Lisbon and then to London to attend a two-day meeting of the Royal Society. I gave a paper on the work that we had been doing in all of the various areas of accelerator mass spectrometry at Rochester over the past couple of years. The meeting was organized by Ted Litherland, Ken Allen and Teddy Hall. We stayed at a hotel in London that was within reasonably easy walking distance of Buckingham Palace, the Marble Arch and the headquarters of the Royal Society. The weather in London was absolutely magnificent.

The first day in London, we had lunch with David Sox at a restaurant near our hotel. This was the first opportunity I had had to talk to him directly after a number of years. He hadn't changed very much—still as dapper and unpriestlike as ever. He was very interested in the events that had led to postponement of the workshop. He had no insights into the reason for the postponement but was inclined to ascribe it to opposition by the Turin sindonology cabal.

On the evening of the first day of the Royal Society meeting, a reception was held at the Goldsmith Hall in London. Teddy Hall was then the Prime Warden of Goldsmith Hall. That organization is concerned with preserving the purity of gold coinage in Britain—ensuring that there is no debasement of the metal. There was an excellent exhibit of the history of British gold coinage and the role Goldsmith Hall played in maintaining its standards.

While we were in Portugal, I had received several messages saying that Professor Chagas would be in London staying at Brazil House during the time we would be there. I phoned him and arranged for a meeting of those people attending the Royal Society meeting who would have been involved in the postponed workshop. Chagas suggested that we meet at Brazil House after the Royal Society meeting ended on Friday 27 June at 5:30 in the afternoon.

Hall had his Mercedes-Benz convertible parked outside the Royal Society headquarters. It was a magnificent vintage automobile. The top was down and the weather was beautiful. He offered to drive us to Brazil House from the Royal Society. Shirley sat beside him in the front seat and I sat in the back with Paul Damon and Doug Donahue. He drove down the Mall, through Marble Arch, by Buckingham Palace in an incredible stream of traffic—Teddy drives with considerable verve. We arrived at Brazil House and went inside. Chagas had been out and arrived a little later. The people at this meeting were Doug Donahue, Paul Damon, Teddy Hall, Robert Hedges, Robert Otlet, Nick Conard (a University of Rochester graduate student who worked in my laboratory), Shirley Brignall and me.

Chagas was quite frank and open about the reasons for the meeting being postponed. They had to do with his interview with Peter Jennings, but also with who were to be invited to the workshop, how the meeting was to be organized, who was going to chair it and where it would be located, Rome or Turin. He told us that he would be sending out a new letter of invitation to the workshop which would be held in Turin, 29 September to 1 October.

On Friday evening there was a banquet hosted by the Royal Society at their headquarters and all of the speakers and their guests were invited to attend. One of the guests at the dinner was Dr P T Warren, Secretary of the Royal Society. After the dinner, he led us to the third floor into the office of the President of the Society. We were shown the official membership book which one signs when one is inducted into the Royal Society. It has pages that go back to the origins of the Society. In particular there were illuminated pages that commemorated each change in succession to the throne in Britain. The British monarch is the head of the Royal Society, and each time a new monarch is enthroned a page is added to the Royal Society book. We were shown signatures of such luminaries as Samuel Pepys and Sir Isaac Newton among others. It was a thrilling experience to see the original signatures of such famous people so many years in the past.

Following the Royal Society meeting, there was a workshop on accelerator mass spectrometry at Oxford University. The accelerator in Hall's laboratory was a Tandetron designed and built by Purser's company, General Ionex Corporation. The facility was run by Robert Hedges and had been in operation now for several years. It was exclusively devoted to carbon dating.

The workshop started on 30 June and I was scheduled to give one of the first talks—an update on dating the shroud. I gave a brief account of the present situation. I mentioned the interlaboratory comparison, the results of which had been reported by the British Museum in Trondheim in June of 1985, the informal meeting we held there, out of which the Trondheim Protocol resulted, and how Carlos Chagas, President of the Pontifical Academy of Sciences, became involved. I described the plans that had been made for the workshop in Turin in June to be chaired by Chagas and to be held in the Archbishop's Palace. Then the article by Peter Jennings appeared in *Our Sunday Visitor* and shortly after that, the meeting was postponed. I described the meeting just held with Professor Chagas at Brazil House in London, with representatives from Arizona, Harwell, Oxford and Rochester. There Chagas announced that the workshop would take place at a time soon to be announced. Chagas stated that he had approval from both the Vatican and Turin to conduct this workshop.

I said that some of the matters to be discussed would be: (1) the sample size required, and the procedure for removing samples to be supervised by a textile expert from Bern, (2) whether the measurements would be blind, (3) assuring that the British Museum would be involved in analysing the data, and (4) the procedures to be followed in due course concerning announcing the results to the public.

At the end of the workshop in Oxford, I told the delegates that May 1987 would be the 10th anniversary of the measurement of carbon-14 in natural samples at the University of Rochester and at McMaster University. We were seriously considering holding the fourth AMS symposium at Rochester in the spring of 1987. I was asked to convey the thanks of the delegates to the organizers of the workshop, which I did.

When we returned to Rochester, both Shirley and I received an invitation dated 2 July 1986. It read as follows:

"Having been directed by the Holy See on 23 December 1985, to organize in collaboration with His Eminence Cardinal Ballestrero of Turin, a preliminary meeting of scientists and specialists to discuss the problems concerning the carbon dating of the Shroud of Turin, I hereby invite you on behalf of Cardinal Ballestrero and myself to attend such a meeting to be held in Turin September 29 through October 1, 1986.

"It is His Eminence's desire and mine also, that during this meeting an agreement be reached regarding the number of samples to be taken, their sizes and their distribution to the laboratories that will perform the measurements.

"We hope that a full debate on the methodology will take place and that all the problems of possible contamination, as well as the possibility of using charred sections of the Shroud, will be taken into consideration.

"While both the Holy See and the Cardinal are committed to carry out the C-14 dating, the final decision to proceed, based solely on criteria of preservation of the Shroud, is up to the Cardinal as Custodian of the integrity of the Shroud itself.

"I count on your keeping this invitation strictly confidential as this is required for the complete success of this difficult enterprise.

"Should your Institution be unable to take care of your travel and living expenses, please let the Pontifical Academy of Sciences, Vatican City, know at your earliest convenience [a telephone and telex number was included]. Sincerely yours, Carlos Chagas."

It had an interesting postscript that said "Hotel reservations and the religious institution hosting the meeting will be communicated to you by Professor Luigi Gonella". Canuto persuaded Chagas to add the postscript to ensure that the meeting did not take place at Turin Polytechnic.

On 14 July I phoned Canuto. Canuto began by saying (modestly) that he could only credit himself with saving the situation. He said a memo had been sent to the Vatican from Turin accusing Chagas of violation of confidentiality because of the Jennings interview and making other baseless charges. It was this communication that triggered the postponement. Gonella came to Rome and met with Chagas and Canuto to discuss the reconstituted workshop. Gonella suggested the wording of a letter that should be sent to all the previous recipients. That letter would state, among other things, that, after the workshop, only two of six laboratories would be selected to do the dating, that scorched cloth would be used and that the Turin Egyptian Museum would analyse the data. He also said that there must be invitations to people other than the ones that were originally invited, namely Jacques Evin from Lyon, a person from the J P Getty Museum and Alan Adler. Canuto said that Chagas seemed prepared to accept all this, but that evening Canuto convinced him that such a letter was ridiculous. Together they composed the actual invitation letter. Canuto said that every word of this letter of invitation was carefully considered. Canuto made a translation into Italian and took both the English and the Italian version with a cover letter from Chagas to Turin to deliver in person to Cardinal Ballestrero. Ballestrero then called the Vatican a day or so later and said that he agreed with the wording of the invitation letter.

On 22 July I replied to Chagas' workshop invitation. I said I had talked to Canuto on 14 July and described to him the meeting we had with Chagas at Brazil House. Our consensus was that the workshop would proceed this time. However, since the main reason for the postponement was Professor Gonella's objection to the relatively innocuous article by Peter Jennings, I thought I should mention that Professor Hall had showed me some correspondence he had received from Ian Wilson. I noted that Wilson is chairman of the British Turin Shroud Society. He is a well known freelance reporter and the author of a couple of books on the shroud. I said it was clear from the material that Hall showed me that Professor Gonella was keeping Wilson fully informed about major details of the workshop, including the exchange of cables which signalled its postponement and of Gonella's meeting with Chagas at the Vatican. Wilson in turn passed this information on to a large and varied group of people in his British Shroud Society distribution list.

Included in Wilson's letter of 24 June 1986 to Hall were three memos from Wilson, marked confidential, to 14 people—mostly in Britain but also to Gonella and D'Muhala. The first of these stated that Gonella was giving STURP the green light for carrying out a battery of tests in July. The second said that Gonella phoned from Turin that, due to complications relating to radiocarbon dating and some logistical problems on the part of the US STURP group, the planned testing was postponed. In the third memo, dated 20 June 1986, Wilson said there was no reason given for postponing the 9–11 June 1986 workshop Chagas had organized to plan for carbon dating the shroud. In his letter to Hall, Wilson blamed Chagas for the postponement and said he found it difficult to understand why, or how long he can keep stalling.

I said that this struck me as a serious breach of confidence and was remarkable evidence of double standards on Gonella's part. For him to be so inordinately critical of Chagas for his interview with Jennings and then to provide all sorts of sensitive information to Ian Wilson, who wrote for the press as did Jennings, was, in my view, inexcusable. To his credit Wilson never did make the information he received from Gonella public.

In my letter to Chagas, I said that a substantial amount of effort had to be devoted to planning details of the workshop and that an agenda had to be agreed upon by the seven institutions. I offered to prepare one. I ended the letter by accepting the invitation to participate.

On 24 July I spoke to Canuto. He said that on 15 August he was going to Brazil for a week. He suggested that I call Chagas to get his verbal authorization for me to proceed to inject some organization into the workshop. If he agreed, I should then prepare a draft agenda for the meeting to be sent for comment by the participants and give a copy to Canuto to take to Brazil.

I talked to Canuto again on 25 July. He suggested that Shirley and I come to New York. Ambassador Celli would be back on 4 August. Canuto would like us to meet his wife and to have discussions with respect to the Turin workshop with Celli included. He said that he was guessing that Gonella would be working up an agenda although he was not asked to do so by Chagas. At the Chagas/Canuto/Gonella meeting in the Vatican in June, Gonella kept raising technical questions. Chagas responded that these were matters to be discussed at the workshop. This may have given Gonella incentive to go back to Turin and prepare an agenda.

On Saturday, 26 July, I phoned Chagas in Rio. I asked him if he had received my letter of 22 July and he said that it had arrived yesterday. I mentioned that I had been discussing the question of an agenda with Canuto, and that Canuto had suggested I give him a call. He said "There is an old proverb [perhaps Brazilian] that says the dogs bark, but the caravan moves on". I laughed as if I knew what the hell he was talking about. He said he would be in Rome by the middle of September. The last letter that he had from Rovasenda said that everything was still on track. I said that I would like to start working on an agenda for the meeting in consultation with the other laboratories. I said I wanted his agreement that I do so. He said, "You just make it [the agenda] and send it to me. I'll send a letter to all the people saying I asked you to do it." I said that I would be meeting with Canuto during the week of 3 August. I would bring him a version of the agenda which we would discuss with Ambassador Celli. Canuto would be visiting Brazil in August and he could bring this agenda to Chagas then. I said that I was now quite hopeful that things are going to work this time. He said, "I'm absolutely sure it is going to work". I said I thought we really had to go into it well prepared with what we were going to talk about and Chagas said "that's certainly true". I said I would prepare the agenda and would consult fully with all the other laboratories. I asked him if he would be good enough to send me a list of who was going to be at the meeting, and he said it was the same list as before with two additions. Druzik and Adler. I said that we had all really enjoyed meeting him in London and he said, "A very happy circumstance I would say".

I talked to Canuto again on 28 July and told him of my phone call to Chagas. I said Shirley and I would meet him in New York on Thursday 7 August. He said he would call me on Monday or Tuesday with respect to Ambassador Celli's availability. I spent the next few days preparing an agenda for the meeting to bring to New York for discussion with Canuto and Celli.

I received a letter dated 4 August from Professor Chagas from Rio de Janeiro. He began by thanking me for my phone call and the letter of 22 July. He said that he had spoken to Canuto and that he knew I was going to discuss an agenda with him that would be sent to Rio. He said he was sure that we were going to meet some difficulties, but had no doubts that we were going to succeed. He again quoted the saying 'The dogs bark, but the caravan goes on'. This is a familiar old Brazilian proverb apparently. He said that everything would be all right in Turin.

Shirley said she had been trying to figure out a circumstance in which she could use this proverb—so far without success!

Chagas continued that he knew one had to be very careful of some people around the Cardinal, but he was sure that he would prevail since he had clear support from His Holiness himself. He said he need not tell me that he was going to chair the meeting. He had some experience in chairing very difficult meetings. He remembered some of the more than a hundred meetings he had chaired either in Rome or in Brazil or at the United Nations, and with patience and calm he always overcame any obstacles.

On 6 August 1986, Shirley and I flew to New York City and

the next day we toured the recently refurbished Statue of Liberty. That evening we had dinner at Canuto's house, a dinner that Vittorio cooked. Monsignor Celli, the Vatican ambassador to the United Nations, was present and of course Canuto's wife. It was an extremely pleasant evening. We had a chance to go over the agenda for the Turin workshop.

In discussing the agenda I had prepared with Vittorio Canuto and Monsignor Celli, there were some relatively minor changes made. It was suggested that in the letter to Chagas, in which this proposed agenda would be enclosed, I state that the agenda had been shown to both Monsignor Celli and Canuto. Celli suggested also that I recommend to Chagas that he show the agenda to both the Vatican secretary of state and the pope.

On 8 August I wrote to Chagas enclosing the proposed agenda. I noted that it had been discussed with Celli and Canuto. I went on to say that if the agenda were satisfactory to him, I would like copies to be sent to a senior representative of each of the six laboratories, the British Museum and to the textile expert, Madame Flury-Lemberg. Comments from them should be requested so that an agreed upon final document would be available before the workshop took place. I would leave it to him whether it should be sent to the other participants. I also noted the suggestion Celli made that copies be sent to the Secretary of State, Cardinal Casaroli, and his Holiness the Pope.

On 26 August I phoned Chagas and his secretary told me that he had been trying to call me. He came to the phone and said that he had received the agenda and he agreed completely with it, but he proposed what he considered a minor change. He suggested that instead of referring to six laboratories, we simply refer to them as the carbon dating laboratories—the word six should not be used. I should have become instantly suspicious of this change. In retrospect, I now realize that Chagas was trying to avoid controversy by eliminating the phrase 'six laboratories' and substituting 'carbon dating laboratories'. He realized that Gonella was opposed to a large number of labs being involved. It probably would not have been harmful to have been unspecific about the number. Nonetheless, I pressed the point about specifying six labs—an action I now realize was unwise.

He said that he was going to send the agenda to everyone that very day. He would not send it to Casaroli, but to the Undersecretary of State Cardinal Somolo, who actually knew more about the workshop and also had frequent access to the pope. He said that Adler of Western Connecticut State College was invited. I asked him whether he was going to ask recipients of the agenda to send him any suggested changes and he said no, this is *the agenda*. He told me that on 20 September he would be flying to Rome and I said Shirley and I would also be in Rome prior to the Turin workshop.

The next day I phoned Chagas and told him that I was sending him a letter concerning the changes that he suggested in the agenda in his 26 August phone call and he said he would delay in sending out the agenda until he had received my letter. He said that there would be 23 participants, including the rector of the Pontifical Academy of Sciences, the vice rector and also Adler.

In my letter to Chagas dated 27 August I said that in the phone call of 26 August he had stated that he would be sending out the agenda immediately and that he had suggested some changes. I said that I was afraid that I had been caught a bit off guard and acquiesced too quickly during our phone conversation. In particular, I wanted to argue that the word 'six' before laboratories remain. I went on to say that if, in Turin, the ecclesiastic authorities insisted on either (a) limiting the sample size to such a small amount as to be insufficient for all six laboratories, or (b) deciding that some of the laboratories are less meritorious, or both, the question should be settled at the workshop. Six laboratories participated in the British Museum test, so it seems very natural and unexceptionable to use the word six in the agenda document. I also suggested that he might somewhere indicate on the agenda that I had been involved in its preparation. I confess that I wanted some explicit recognition from Chagas that I had put considerable effort into organizing the workshop.

I concluded by saying that I realized that acceding to my request to include the number six might raise the barking level, but there was a saying in English, 'barking dogs seldom bite', and I also gave him the French translation of that particular saying 'Tout chien qui aboye ne mord pas'.

Shortly after this, I received a letter from David Sox dated 29 August saying that he had received a letter from Peter Rinaldi that the carbon-14 meeting would occur at the end of September in Turin. Sox said that Richard Burleigh was now at the British Museum of Mankind.

On 2 September I phoned Chagas and asked him if he had received my 27 August letter. He said that he had. He then went on to say that he wondered if he could be frank with me. He said that he didn't know me very well nor I him, but he had the impression that I did not trust him. Perhaps that was because he was a Latin-American, or in a different field of science, or for some other reason. He said he did, after all, have some experience with organizing meetings.

I must say that his statement that I didn't trust him surprised me. I certainly worried from time to time as to how influential he was in the rarefied atmosphere of Vatican politics and even whether he was a match for Gonella. The fact that Chagas was a Latin-American was absolutely irrelevant as far as I was concerned. I had the highest respect for him as a scientist. I now realize he would not have raised the question of trust had I immediately agreed to the elimination of the word 'six', and it probably would not have mattered if I had.

He went on to say he still wanted the word six removed from the agenda. He does think that all six should participate in the carbon dating, but that question should be settled in Turin. So I finally acceded. He also said that he wanted to chair the last two sessions, the one on public disclosure of results, and conclusions. In the agenda I had prepared, I had various of the principal participants listed as moderators for the various topics to be covered. This change would eliminate the two people I had assigned to these topics—one of them being Gonella. Chagas said he would add some words like 'the Pontifical Academy of Sciences has asked Professor Gove to organize the workshop and is grateful for his efforts'.

As a result of this phone call, I drafted a letter to Chagas that expressed some of the feelings I had about the shroud dating enterprise and about the role that Professor Gonella was playing. I said that I was not used to getting this heavily involved in an activity that fails because of extraneous and irrelevant machinations of individuals with highly questionable motives. I said that Chagas' perception that I lacked trust in his leadership ability and in his standing as a scientist because he was a Latin-American or in a different scientific field from me or for any other reason was just simply wrong. I went on to say that maybe the real problem was that he lacked trust in me and the kind of role that I had been playing in this operation. If that were the case, perhaps it would be best for the whole enterprise if I would just withdraw completely.

As I usually do, I showed a copy of this letter to Shirley to get her reaction to it and she urged me *not to send it* until I had discussed the letter with Canuto. I learned that Canuto was vacationing at Lake Maggiore in northern Italy. I finally contacted him on 3 September. I read the letter to him. He said that he did not understand what Chagas was saying, but he urged me *not to send the letter offering to withdraw.* That would just play into Gonella's hands. He said he would be returning to New York on Sunday and would call Don Carlo, his name for Chagas, on Monday. As a result of this, I decided not to send the letter. I did however send a copy of the letter to Vittorio on 4 September. I said that the draft of this proposed letter pretty well sums up my position. I said that although I would not send the letter, as a result of his forceful pleading, I was very distressed by Professor Chagas' frank comments. I thought that his assumptions about my opinion of him set a very bad tone for the workshop. In my view, it reflected a lack of trust by Chagas in me. If that were so I felt that my withdrawal from the enterprise would increase its chances for success.

On 5 September Chagas called me. He did not say that Canuto had talked to him but it was obvious that he had. Chagas said that he had hoped that his private frank comments to me would not be taken too seriously. He said that he had the highest regard for me as a scientist, and that I was more key to the success of the shroud enterprise than he was. He said that it would fail if I did not participate. He was putting a phrase into the document that the Academy had requested me to organize the workshop. He asked whether we would be in Rome prior to Turin and if so we should plan to meet with him. I said that we would be there, and that I would contact him when we arrived. He said he wanted to mail the agenda with a cover letter that day and again he wanted to chair a couple of the sessions he had mentioned previously. I was quite pleased with this phone call.

I really did at that time, and still do, have enormous admiration and indeed affection for Chagas. I recognized that, in dealing with people in Turin, he was up against a problem that few of us have ever had to face in our lifetime. He was probably doing as well as or better than anyone else could under the circumstances.

A few days later Shirley and I received the agenda, marked confidential, with copy numbers on each. There was also a list of the participants who were to attend the workshop. In alphabetical order after Chagas' name they were: Alan Adler, Western Connecticut State University; Shirley Brignall, University of Rochester; Vittorio Canuto, NASA; Paul Damon, University of Arizona; Robert Dinegar, Los Alamos National Laboratory; Doug Donahue, University of Arizona; Jean-Claude Duplessy, Gif-sur-Yvette (a new name—someone I had never heard of); Jacques Evin, Lyon; Mechthild Flury-Lemberg, Abegg-Stiftung; Luigi Gonella, Turin Polytechnic; Harry Gove, University of Rochester; Teddy Hall, Oxford; Garman Harbottle, Brookhaven National Laboratory; Robert Hedges, Oxford; Steve Lukasik, J P Getty Conservation Institute (this was an incorrect address); William Meacham, Hong Kong; Robert Otlet, Harwell; Enrico di Rovasenda, Director Pontifical Academy of Sciences; Mike Tite, British Museum; and Willy Woelfli, Federal Institute of Technology, Zurich.

On 23 September Canuto called. He said that Gonella had called Dardozzi, the vice chancellor of the Pontifical Academy of Sciences, on 17 September and had a one and one half hour conversation with him. Gonella, in Canuto's phrase, had been blown out of the water by the following: (1) that the Pontifical Academy of Sciences had thanked me for organizing the meeting. Canuto said that they should also have acknowledged the fact that I was the principal investigator on the National Science Foundation grant that saved the Academy some $6000 in travel costs. (2) Gonella was disturbed that Turin had not been consulted on the agenda. Canuto said that the involvement of Monsignor Celli in deciding on the agenda would prevent that criticism from being valid, but I pointed out that Celli was from the Vatican and thus suspect to Turin. I also pointed out that I had provided Gonella with an agenda, essentially the same as the final one, when he met with us in New York. (3) Gonella was bothered by the fact that he was not listed anywhere as a moderator of any session. Gonella felt that this was a real slight. At the actual workshop, the question of moderators was academic. Harbottle suggested that Professor Chagas chair the whole thing. That particularly amused me because of the often and forcefully expressed view by Harbottle that getting the Pontifical Academy of Sciences involved in the first place was a really bad mistake on my part.

Chapter 6

The Turin Workshop

On 23 September 1986, Shirley and I left for Rome. I phoned Chagas and he invited us to visit him at the Pontifical Academy headquarters in Vatican City on 25 September at 11 am. We did so and were taken to Chagas' office—he arrived a few minutes later. A meeting organized by the Academy was being held at the time involving some 20 or 30 delegates. It concerned the interaction between the atmosphere and the oceans and its effect on weather.

During our meeting, Professor Chagas combined reminiscences of his early days with remarks about the forthcoming workshop and other matters. He started studying medicine at the age of 15 and became a full professor at 26. He collaborated with Robley D Evans on the construction of counters for the detection of radiophosphorous and radioiodine. Evans had been a professor of mine at MIT when I was studying for my PhD in the late 1940s. Chagas said that Gonella's problems were with him, not with me. He noted that Ballestrero was not convinced of the shroud's authenticity and he (Chagas) wondered why none of the Savoy kings had ordered any tests to be made on the shroud, especially since two out of the four were anti-clerical. I asked him about the 'blood' on the shroud and its analysis by Baima Bollone and Alan Adler. He said there was some evidence that it was primate blood. I asked him as an MD and a biochemist if he thought the evidence was definitive and he said he did not. This would come as a shock to Adler and others who seemed absolutely convinced it was. We also talked about the possibility of samples being taken during the workshop and whether charred material could be used.

We had lunch with the delegates to the meeting and sat at a table with Chagas. After lunch the delegates, along with Shirley and I, were given a marvellous tour of the Vatican Museum. The

141

next day we were invited to participate with the delegates in an audience with the pope, John Paul II.

We arrived at the Academy at the appointed time and were taken by bus to the place in the Palace where these audiences were held. We were seated before an elegant throne flanked by two lesser throne-like chairs. The pope arrived accompanied by two cardinals. Members of the Swiss Guard in their colourful uniforms of blue, red and yellow vertical stripes and their large floppy black berets saluted. The Swiss Guard really did comprise well turned out and uniformly handsome young men of Swiss nationality.

When the pope was seated, Chagas made a short speech telling him what the meeting was about. The pope then read a speech of welcome in English, probably written for him by someone at the Academy. His English was excellent, even to the extent that when he once mispronounced a word, he quickly corrected himself. Each delegate then introduced himself or herself to the pope, had a photograph taken shaking hands with him, and returned to their seat. We were personally introduced to the pope by Professor Chagas, first Shirley, her knees shaking slightly, and then me. We later received really excellent photographs of this exciting occasion. Needless to say, I had no discussions of dating the shroud with the pope but he knew from Chagas' introduction that we were in Italy to attend the Turin workshop on that subject. We declined Chagas' invitation to again join them for lunch, partly not to take too much advantage of his hospitality and partly to avoid another formidable Italian lunch. During our stay in Rome we saw virtually every important historical and tourist attraction. It is a city of such immense grandeur as to be overwhelming.

We checked out of our hotel in Rome on Sunday 28 September and arrived in Turin late that afternoon. Waiting for us in our hotel was an ornate folder with information on the workshop, provided by Gonella. It contained a map of how to reach the seminary where the meeting was to be held and a list of participants. It was the same list Chagas had provided us. It also contained the information that "The meeting is held under the responsibility of His Eminence Anastasio Cardinal Ballestrero, Archbishop of Turin in his quality [sic] of Custodian of the Holy Shroud and for the scientific part will be chaired by Professor Carlos Chagas, President of the Pontifical Academy of Sciences. The meeting will be held at the Seminario Metropolitano of Turin 83 Via XX Settembre." The opening session would be 29 September at 9:30.

At the entrance to the Cathedral of St John the Baptist: Peter Rinaldi and Shirley Brignall.

The folder also included the English translation of a telegram from Professor Chagas to Cardinal Ballestrero that read, "With reference forthcoming Turin meeting on Holy Shroud I want to assure your Eminence that the agenda recently mailed to participants has proposing character as usual, and is to be examined in collegiate way. With the occasion I ask you to accept devoted obsequious feelings, Carlos Chagas, President of the Academy of Sciences." Included were some black and white photographs of the shroud and an English translation of the short address that Cardinal Ballestrero was to give the next day at the opening of the meeting.

On Monday, 29 September 1986, Shirley and I walked from our hotel to the Seminario, the site of the Turin workshop. The meeting was held in a building that formerly served as a seminary for young men studying for the priesthood. It was now practically empty, since fewer young men pursue that career. Those that do are housed in a smaller establishment outside Turin.

On the way in, I introduced Shirley to Father Peter Rinaldi, who captivated her as he does everyone who meets him. He first experienced the fascination of the Shroud of Turin as an altar boy in the cathedral adjoining the chapel in which it was stored. For many years, he was a parish priest in Port Chester, New York.

We entered the meeting room and seated ourselves as we wished around tables arranged in an open rectangle. Chagas was seated at the head set of tables, flanked by Rovasenda and Gonella

Harry Gove, Jean-Claude Duplessy (Gif-sur-Yvette), Peter Rinaldi and Alan Adler (Western Connecticut State University) at the cathedral entrance.

to his right and Canuto to his left. I noted that the aggressive Dinegar had snared a prime position to the left of Canuto, so he was at the head tables, whatever that was worth. At the tables forming one long side of the rectangle on the right sat Dardozzi (just around from Gonella), Shirley, me, Woelfli and Damon. On the long side to the left sat Lukasik (just around from Dinegar), Donahue, Flury-Lemberg, Harbottle, Duplessy, Hedges and Tite. On the short end opposite Chagas, from his left to right, sat Hall, Evin, Otlet, Adler and Meacham. This seating arrangement was more or less maintained during the entire three days.

An elaborate recording system operated by a technician filled one corner of the room with inputs from several microphones around the table. The entire proceedings have since been transcribed—471 pages in all. It was remarkably well done despite the fact that it was transcribed from English, the language used by everyone at the workshop, by a secretary whose native tongue probably was Italian. What follows is based on notes Shirley and I took during the proceedings, supplemented by material from the official transcript.

At around 9:30, His Eminence Archbishop Anastasio Cardinal Ballestrero arrived. He looked older and greyer than when I had last seen him eight years ago in 1978—but then so did I. He took Rovasenda's place at the head table and Father Rinaldi stood

Official opening of the Turin workshop, 29 September to 1 October 1986. Carlos Chagas (president of the Pontifical Academy of Sciences), Cardinal Ballestrero (Archbishop of Turin), Reverend Enrico di Rovesenda (director of the Pontifical Academy of Sciences), Robert Dinegar (Los Alamos National Laboratory chemist, bomb designer, Episcopalian priest and chairman of STURP's shroud carbon dating committee). Giovanni Riggi (Turin microanalyst), standing.

behind him. We introduced ourselves to the Cardinal starting with Shirley.

Ballestrero then gave his opening statement in Italian. The English text follows:

"As my specific competence is not in a field appropriate to an active participation in the forthcoming discussions I shall not take part in your talks, leaving to my assistant, Professor Gonella, the task of liaison with me.

"I hope that these days of study will bring out such conclusions to allow presenting a valid and acceptable project for at last carrying out the radiocarbon dating of the Shroud cloth, a test that, owing to the uniqueness and singular character of the object, certainly could not easily be repeated.

"The project coming out of this meeting, including concrete operative proposals, will be submitted to the Higher Authority of the Holy See, as it is explicitly requested in the letter sent to me on 24 September by the Secretary of State.

"And as a final thought at this point, it is only left to me to wish you all a serene and fruitful work".

Robert Otlet (AERE Harwell), Robert Hedges (Oxford), Doug Donahue (Arizona), Harry Gove (Rochester).

Then followed the opening address by Professor Chagas. He paid tribute to Cardinal Ballestrero, whom he described as having been instrumental in fostering the understanding of the mystery surrounding the shroud and its history. He then gave a brief historical summary of the events that had taken place with respect to scientific testing of the shroud. He expressed his sincere appreciation to the Shroud of Turin Research Project (STURP) for their scientific contributions to understanding the shroud.

He said that the only aim of this meeting was to establish a protocol for the radiocarbon dating of the shroud. A preliminary agenda of the important items to be discussed had been prepared with the help of Professor H E Gove, whom he wished to thank. This agenda, like that for any meeting, could be changed before the workshop began. Regrettably he had not been able to mail it to the delegates in a timely fashion. He thanked Dr Vittorio Canuto who had given him valuable assistance. Thanks were also due to Professor Luigi Gonella for his valuable collaboration with the Pontifical Academy of Sciences in organizing this workshop.

In the agenda I had prepared, referred to above by Chagas, I had designated people to serve as moderators or chairs of various sections of the workshop. Harbottle now suggested that Professor Chagas serve as chairman during the complete workshop. This was a remarkable change in the opinion he had expressed to me

Father Peter Rinaldi (Turin) and Harry Gove (Rochester).

that it was a mistake to get the Pontifical Academy involved in the first place. I am always willing to accept converts to my point of view, however, and his suggestion was accepted unanimously.

At a point later on in the workshop, a list of the participants was passed around for people to correct their addresses or make whatever changes they wanted. Several people did make changes, for example, Professor Alan D Adler of the Chemistry Department, Western Connecticut State University, removed his name completely. Professor [sic] Vittorio Canuto put as his address USA, omitting the NASA connection. Dr Robert Dinegar gave as his address USA Shroud of Turin Research Project only, with no location. Finally, the person listed as Professor Steve Lukasik, from the J P Getty Conservation Institute, corrected his name to Dr Stephen J Lukasik and gave as his address Shroud of Turin Research Project, USA, just as Dinegar had done. The name of a representative of a seventh carbon dating laboratory had been added to the list, Jean-Claude Duplessy, of the AMS laboratory at Gif-sur-Yvette in France.

I was a little surprised at the requested changes. Perhaps they reflected a slight embarrassment at indicating a connection between such esteemed organizations such as NASA (Canuto), Los Alamos National Laboratory (Dinegar) and Northrup (Lukasik) and the Turin Shroud. In the case of Adler, I suspected it was a case of playing hookey.

Luigi Gonella (science advisor to Cardinal Ballestrero), Carlos Chagas (president of the Pontifical Academy of Sciences) and Vittorio Canuto (NASA).

Shirley and I had lunch with Lukasik on the first day, and he gave me his card. He is Vice President–Technology of the Northrup Corporation whose headquarters are in Los Angeles, California. Northrup is a manufacturer, among other things, of military aircraft, and supplies one of the fighter planes for the US Air Force. It is interesting that there are several 'warriors', including atom bomb builders from Los Alamos, connected with STURP. Perhaps it is some sort of compensation for their occupation that makes them take an interest in the 'Prince of Peace'.

Gonella then gave a long, rambling talk—it occupies almost 30 pages of the official record. He described how he became involved as chief shroud scientist. The cardinal asked Turin Polytechnic to suggest someone to advise him on science matters and his colleagues elected him. His credentials were that he presently held the chair of physical measurements at Turin Polytechnic and is in charge of the curriculum for nuclear engineers. I later discovered why the cardinal had not requested suggestions for this post from the much more prestigious University of Turin—their faculty turns out to be much more left-wing.

Gonella's main concern was to make sure that the people asking to make tests or to do research on the shroud were not crackpots. "I can assure you that the shroud attracts crackpots as honey the flies." He gave his views as to why people consider carbon dating to have a special status compared to other scientific tests and why many think it should be decoupled from them. It was not made clear whether he agreed.

He went on to say that in February 1979, Gove *et al* and Harbottle *et al* were the first group in recent years to offer to date the shroud if Turin desired. The offer got sidetracked but was resubmitted in August 1979. Later this proposal was backed by STURP. Gonella said this proposal suggested using the 1973 Raes sample, but Turin discovered that the chain of evidence authenticating the Raes samples was irremediably broken. Raes showed off the samples in his study as a conversation piece. He later mailed them back to Turin. There were two samples, one from the edge of the main body of the cloth and one from the side strip. When Raes received them they were stitched together but he unstitched them. When he sent them back he failed to identify which was which. Gonella admitted that it was obvious by looking at the place on the shroud from which they were removed which was which, but it would not stand up in a court of law. He did not elaborate on why it would not.

He then stated that in 1943, because Turin was near the French border and there was fear of bombing, the shroud was sent to Montevergine. When it was returned to Turin in 1973, the cardinal wanted to have a look at it—to make sure it was undamaged. I thought to myself, was Gonella claiming that the shroud was away from Turin for some thirty years? Perhaps the archbishop of the time just forgot about it as he forgot about the Raes samples until someone jogged his memory. Gonella must have been confused, however, because, as stated in Chapter 1, the shroud was moved to Montevergine in 1939 and returned to Turin in 1946. In any case, Gonella said that a commission was appointed and it was this commission, in 1973, that permitted the shroud to be displayed for TV and press purposes and allowed Raes to take his samples. Gonella said that Raes had taken approximately 2 square centimetres from the main cloth plus approximately 17 cm of thread. He admitted that the amount of information that was obtained by Raes was minimal. In any case, he said there was no point in using the Raes samples for carbon dating. I found this surprising. It was the first time Gonella had stated that the Raes samples should not be used. Gonella himself had suggested their use for carbon dating at the STURP workshop in Santa Barbara in 1979. When we offered to date the shroud in 1978 we had no idea that such incredibly sloppy control had been exercised over the Raes samples by the archbishop of Turin—almost as if they had been cut from a dishrag. I now agreed it would be best to no longer consider using them for carbon dating.

He went on to say that until the death of the former king of Italy, Umberto II, the Shroud of Turin was the private property of the House of Savoy although it was confiscated by the Italian Government when Umberto was banished after World War II. It was stored in Turin in a chapel which, although connected to the Cathedral, was now owned by the state and not the Church. The Bishop of Turin was designated by the state as its custodian. When Umberto willed the shroud to the Vatican, legal questions arose as to whether it was actually owned by the state (due to the king's banishment) or by the king. This was settled in the courts and now the shroud officially belongs to the Holy See and the Archbishop of Turin continues as its custodian. There are fears in Turin, however, about its ultimate place of residence. The Shroud came to Turin in the 16th century 'provisionally', so the Turinese are uneasy whenever it is suggested that the shroud go to Rome 'provisionally'.

Gonella went on to say that the first formal proposal to date the shroud was that of STURP in October 1984. I objected and noted that the first offer to date the shroud had been made by me in Turin in 1978 and then made formal in a letter to Ballestrero in 1979. Gonella brushed the objection aside by saying the 1979 proposal was for the use of the Raes samples. He seemed to have conveniently forgotten that he suggested their use for carbon dating the shroud at that time. The STURP proposal was to take new samples and to make other tests. That started the chain of events culminating in the present meeting. He said he learned through STURP about the 'so-called' Trondheim Protocol of July '85 and its revision of September '85 but it was never formally submitted to Turin and thus did not count! Gonella seemed hell-bent on historical revision.

He explained how a seventh laboratory came to be added as a participant. In April 1986, an independent proposal to date the shroud had been submitted by J-C Duplessy of the Gif-sur-Yvette AMS laboratory which led to his invitation to attend this meeting. Many other tests were proposed by STURP. Gonella said people in Turin were perhaps more interested in the other tests than in carbon dating because they addressed conservation. That, he claimed, was the primary responsibility. (Since Turin had totally ignored the question of conservation for at least 300 years and still do to this very day, that statement by Gonella was incredibly casuistic—in any case, a knowledge of an object's age is the usual prelude to decisions on conservation.)

STURP suggested they should have the responsibility for taking and distributing shroud samples in close collaboration with Turin. The Trondheim–Gove Protocol introduced a new idea. It suggested the use of the British Museum as the coordinating institution. Gonella said it was not clear to him what unique role the British Museum would play. He thought other institutions would do as well.

He commented on blind testing and noted that Turin did not ask for a blind test—it was the scientists who wanted it done that way. He said he did not know how blind testing can be carried out on such a well known object as the shroud. Blind testing was an idea that mainly appealed to laymen. "We trust that the people at your labs do not manufacture data, otherwise it would be rather useless to go on."

There was, however, to Gonella, one cause of great concern. If a shroud sample was required for each of the seven laboratories, an inordinate amount of material would be needed. Should all seven get a sample of the shroud? He felt that we must distinguish between scientific and political motivation. At that point I asked Gonella to define 'political', he said it was anything not 'scientific'. It was clear he considered my motives to be political since, in the letter he wrote to Chagas, he accused me of using the possibility of dating the Turin Shroud as a lever for extracting more support funds from the NSF. I wondered about his personal motivations as much as he did about mine.

It was now Chagas' turn to speak. He explained how the Pontifical Academy had become involved. In October 1984 while he was in Rome, Monsignor Martinez, a Vatican undersecretary of state, gave him a document from STURP proposing a radiocarbon dating of the shroud. He had been asked to evaluate it and to decide how the academy might collaborate. Chagas said he had been pleased to have been asked but noted, in the end, the Trondheim Protocol was better than STURP's. If the shroud must be dated, he emphasized, it must be done right, it must be done now, and this may be the last chance. The meeting agenda was to be confined to the question of carbon dating alone and not, as Gonella seemed to be suggesting, other tests as well.

Hall spoke up. He thought that blind tests were necessary to convince people outside, not because we do not trust each other.

The British Museum is needed as the coordinating institution to make it look convincing. He agreed that seven labs might be too many—maybe five would be enough.

Harbottle said he agreed with Chagas' view that this is a one-time experiment and it must be done right the first time. The number of labs to be involved was an important consideration. In his view two or even three were probably too few.

I emphasized the importance of the involvement of the Pontifical Academy of Sciences. It provided a bridge between science and religion. The Shroud of Turin was a unique object. I did not fear that any protocol we adopted for carbon dating it would set a precedent for other archaeological dating. Dating the shroud should be a remarkable example of the ability of the new technique, both AMS and small counter, to accommodate small samples. With respect to the number of labs that should be involved, there were others in addition to the present seven who did excellent work, e.g. IsoTrace at the University of Toronto. The seven laboratories present here were the only ones that had offered to date the shroud. It was hard to see how one could reduce this number. Finally, I remarked that one would have to have a fertile imagination to suggest political motivation—political as defined by Professor Gonella—that impelled any of the lab people to be involved in the shroud. Maybe there were questions that science could answer, but should not try to answer, and the age of the shroud may be one. Our motivation to do so, if requested, was scientific curiosity and certainly not political.

Tite assured the delegates that the British Museum was not forcing its way in. They would be willing to help if the scientists involved thought that would be useful. If not, they would withdraw with no ill feelings.

There were further comments on the question of blindness, the number of labs to be involved and whether every lab should get a shroud sample. I remarked that the motivation of the people at my lab might be lost if we did not get a piece of the shroud. Otlet said he disagreed—not all labs should get a shroud sample. He was worried about the publicity that any dating of this kind was going to attract. This could be avoided if one did not even know whether or not one was dating the shroud. Evin said he agreed with Otlet that no lab should know whether or not it had a shroud sample.

Lukasik thought that material should be removed from the shroud only to the extent that it could be justified by the information that could be derived from it. He agreed one or two labs dating the cloth were too few—the minimum was three. He did not agree that carbon dating could be decoupled from the other tests. Dinegar said he had spent the last six years of his life on this project. It was really wonderful that we seemed to be coming to a consensus. Adler thought the tests should be blind and that samples from various parts of the shroud should be taken.

Duplessy said he had had many inquiries in France as to why the shroud had not been dated. He was here in response to these inquiries. He was not asking for a sample, but would participate if a decision to date the shroud was made. Hedges said blind samples were tough to deal with, one must have some knowledge of the cloth. The idea of dealing with samples of unknown origin alarmed him quite a bit.

Meacham said that, in his experience as an archaeologist, it was unusual to have so many labs involved. He knew of only one example in which five labs had been involved. In that case they had had ten thousand kilograms of charcoal from the site of some Chinese emperor. (I thought: my God, it sounded like a soft coal mine, not an emperor's tomb—maybe Meacham was a coal miner in disguise.) Anything known about the samples should be known to the daters. Possible contamination should be checked. There could be second measurements—samples could be exchanged (he seemed to have missed the point that with AMS samples are totally consumed). As many zones of the shroud as possible should be dated. (Meacham was here beginning to develop his preposterous analogy of the shroud as an archaeological dig.)

After lunch Chagas asked me to begin the discussion on the size and number of samples. I asked how much the cloth weighed per square centimetre (its areal density). That number and the weight of carbon in a given weight of cloth were of fundamental importance in deciding how much cloth was needed by the AMS and small counter labs.

Gonella responded that the number of threads per centimetre was 26 weft and 38.6 warp with lineal densities of 63 g/km weft and 17 g/km warp according to Raes. (Using these figures one obtains an areal density of 22.9 mg/cm^2.) He said there were estimates of the areal density of 23.7 mg/cm^2 and of 23.4 mg/cm^2 made before 1975. He did not specify how these were arrived at. In 1978, STURP measured a value of 20 ± 5 mg/cm^2 based on the absorption of 12 keV X-rays. Hence it would be reasonable to assume an areal density of 22 ± 2 mg/cm^2.

More importantly, I realized, the lineal density of the warp and weft threads was now known to all the delegates. Unraveling the shroud sample would no longer be sufficient to render the sample blind. It was extremely unlikely that a control sample could be found with warp and weft threads that had the same lineal densities. The question of a blind measurement was now moot. The laboratories would know whether or not they had a shroud sample. The only way around this would be to unravel the samples, clip the warp and weft threads very short and even untwist and separate the individual strands of the threads themselves. No prudent carbon dater would accept such masses of fluff. Cleaning them would be difficult and chancy and it would be too easy for some unscrupulous person along the chain to add sufficient additional fluff to change the age. Neither I nor anyone else at the workshop then or later pointed this out.

I then asked what the conversion factor between cloth weight and carbon weight was? Otlet replied that he had the figures for the three samples that had been supplied by the British Museum for the interlaboratory comparisons. The answer was: 24% by weight of the untreated cloth was solid carbon. That meant, to within $\pm 10\%$, one square centimetre of shroud cloth would yield 5 mg of solid carbon.

Gonella referred to a photograph that he had supplied to each delegate in the packets left at our hotels. It showed that an appreciable area of cloth was covered by the patches applied after the fire in the 16th century. Cutting out these charred and partly charred or scorched areas would not affect the appearance of the shroud and might even be beneficial. Dinegar stated that there were 400 cm^2 of cloth under the patches. (Using all of this would yield 2 grams of carbon—enough to be measured by conventional large proportional counters!)

A long and detailed discussion followed as to whether charred cloth would produce an accurate carbon date. Harbottle briefed us on what he had learned in his survey of many carbon dating labs. The preponderance of the evidence was that it would give an accurate date. To be on the safe side, it was eventually decided that the sample should be taken well away from any scorched or even water marked areas.

Another discussion involved the question of whether the shroud, at one time, had been immersed in boiling oil as some sort of religious test. Adler was firmly of the opinion that it had not been but there is no strong evidence one way or the other.

Meacham then made his archaeological dig analogy—how one takes samples from different layers; samples from different areas of the site; samples from different parts of a hearth—all this gives greater credibility to radiocarbon dating. He urged that the shroud be approached in the same way. I thought to myself that Meacham might make a great lecturer in freshman archaeology but his comments seemed remarkably inappropriate as applied to the shroud.

Gonella again stated it was not a question of the amount of material but the amount of information one can get in the least intrusive way. It turned out he had a hidden agenda that did not become clear until much later—seven labs were far too many.

I then remarked that if we had a better idea of how much material the labs required, some of our discussion would be more effective. If we only needed a total of one square centimetre then we could take a millimetre square from a hundred different spots. That would certainly satisfy Meacham. I asked "Are we dealing with ten square centimetres or 100 square centimetres?"

Flury-Lemberg stated that the cloth is the same from one end to the other. There is no need to take samples from various places. One could take strips from the edges of the main cloth from any place and it would be the same.

Chagas asked me to lead the discussion on how much sample each laboratory needed. I suggested that it would be simplest to ask each lab how many milligrams of carbon they felt they needed—a conversion to cloth area could be made later. If the five accelerator labs each gave a number, we could take an average. I said that a year before Rochester had dated something like 15 micrograms of carbon but the accuracy was poor—that is an example of the sample size going down. For a standard dating like the shroud, 5–10 milligrams of carbon would be needed.

The other AMS lab representatives gave estimates ranging from 1 to 5 mg finally settling on the latter. The small-counter labs at Brookhaven and Harwell requested 10 and 15 mg respectively.

At one point Gonella interjected, "By the way, excuse me, for us to have the whole picture, what is the average activities of your labs?" Chagas said that question could be addressed later.

The final requirements for sample size were a total of 25 mg of carbon for the five AMS labs and a total of 25 mg of carbon for

Paul Damon (Arizona) and Harry Gove (Rochester).

the two counter labs—a total of 50 mg of carbon. That translates
into about 210 mg of cloth—about 10 square centimetres.

Chagas returned to the question that Gonella had posed—
namely the annual activity of each lab—and asked me to respond
first. I asked Professor Gonella why he was interested. I could
give him a complete rundown on our activities, but what was the
point of the question? Gonella said he merely wanted to know
how many carbon dating jobs each lab did per year.

I replied that the NSRL at Rochester was primarily designed
to do experiments in basic nuclear science and operated around
the clock 365 days a year. The AMS group got about one sixth
of the time. We measured a number of cosmogenic radioisotopes
including chlorine-36, iodine-129, and carbon-14. The latter was
not presently a major part of the operation but when occasion
arises to make such measurements, we do so. I could not say
exactly what percentage of the time we spend on carbon-14.

Woelfli estimated that this year his lab would have made
about 1000 carbon-14 measurements along with a smaller number
of measurements of beryllium-10, aluminum-26 and chlorine-36.
Damon said he thought they did about 800 carbon-14 samples per
year. Otlet said his lab did about 600 to 700 carbon-14 samples
a year, of which about 100 were with the small counters. They
also measure tritium and stable carbon isotope ratios. Hedges

said they made about 400 carbon-14 measurements a year on their accelerator. Duplessy said the Orsay Tandetron made about 300 carbon-14 measurements a year. It was used about 60% for carbon-14 and 40% for beryllium-10 and aluminum-26. Harbottle said, being a research lab, they are really not in production on carbon-14. This year they had only done a handful, perhaps 5 or 6, and last year probably as many as 100. Of the seven labs, Zurich, Arizona and Oxford—all using AMS—made the greatest number of carbon-14 measurements per year and they were the three labs Gonella finally selected to do the job.

Chagas then asked whether all seven of the laboratories should receive a shroud sample or whether some should receive control samples only?

Harbottle gave an excellent analysis of why all seven labs should be involved. A hundred things could go wrong—the British Museum interlab comparison tests originally suggested by Bob Otlet had shown that. (Some time back Harbottle had told me it was his idea and "a helluva good one" at that.) This would probably be the last chance to date the shroud and we must do a good job. It was important that both AMS and small-decay-counter techniques be used. Seven labs were better than six or five—four was an impossibly small number (Chagas agreed with that).

Chagas then asked Harbottle whether he thought all seven labs should actually get a shroud sample and Harbottle said he thought six samples to six labs with one getting a dummy sample would suffice. Control samples—probably two per lab—should also be provided (a distinction was being made between control samples and shroud-like dummy samples). He later changed his mind and opted for one shroud sample and one control sample for each of the seven labs. This suggestion had come from Duplessy. Donahue then correctly observed that blindness is increased by using two control and one shroud sample for each lab. (As mentioned above, blindness was impossible because the shroud could be identified even if it were unravelled—but no one spoke up.) Flury-Lemberg flatly stated that it would be impossible to find a shroud-like dummy.

During much of this part of the discussion, Gonella made incoherent teeth grinding sounds from time to time. He was clearly not in favour of seven samples for seven labs. He was also clearly unhappy about the assumption that all seven labs be involved. Harbottle's statement that four labs would be unacceptably too few bothered him greatly. He asked whether this meant that all dating made on only one archaeological sample should be considered unreliable? (I wondered to myself did he really believe the shroud was like any other archaeological object?)

Chagas tried to sum up by saying that we needed two controls from two different age populations plus a shroud sample for each of the seven labs.

Gonella indicated he was very unhappy about this conclusion. Chagas suggested that it was getting late, people were tired and more discussion on this question could be held tomorrow. Gonella would give a slide show at 9 am and the workshop proper would start at 9:30.

On his way into the seminary to start the second day's proceedings, Professor Chagas slipped on the cobblestones at the entrance through the high wall surrounding the building. He scraped his head which bled rather badly and also injured his arm. The cobblestone sidewalk was very narrow and bordered directly on an extremely busy street along which traffic proceeded at a typical Italian clip. He was fortunate not to have fallen into the street and been run over. His wife helped him up. She supplied him with a lovely silk scarf she was wearing—a gift from a dear friend, she told Shirley afterwards—for him to use as a sling for his arm. Despite what must have been very painful injuries, Chagas gallantly carried on.

At 9:20, Gonella's slide show began and lasted for an hour. It concerned the tests on the shroud carried out by STURP in 1978. It showed the shroud racked up on the frame devised by STURP and bathed in bright lights including ultraviolet for photography, for X-ray analysis and for other measurements. His talk had a profound effect on several people present, including Chagas, because it vividly demonstrated how very intrusive the STURP tests were. Whatever the purpose of the show, that was not the one Gonella had planned.

Finally at 10:45, Chagas started the meeting by saying he hoped he could continue as a good chairman despite the blood he had lost from his fall—a lot of it could still be seen on his shirt. He turned to the blood expert, Adler, and said "Perhaps Dr Adler could use part of my shirt to. . .". At least his sense of humour had not deserted him. He said the first conclusion he had reached yesterday was that this was the time for radiocarbon dating the shroud. The techniques that had recently been perfected can give sufficient precision to decide its age.

A large enough sample size was needed for 'perfect' scientific dating. The question of size is connected to the number of labs to be involved. Chagas thought the consensus was seven labs. However, he said, Professor Gonella disagrees with this consensus. Chagas also thought it was agreed that $12\frac{1}{2}$ cm^2 of shroud cloth was enough (this was close enough to the 10 cm^2 to not be questioned) for the seven labs. He said these were the conclusions that were reached and have already been discussed quite thoroughly but he would now like to hear from Professor Gonella his reasons for not accepting this consensus. He remarked that blood loss must have impaired his memory because he should have first thanked Professor Gonella for the slide presentation.

Gonella said that having the seven labs involved as a team was acceptable to him, but did this mean that seven shroud samples were needed? In his view, this was not automatically the case. It is true that $12\frac{1}{2}$ cm^2 was a trivial percentage of the shroud area but percentage had no meaning. Could we use material under the patches? Charred material under the patches was 'waste' material. He thought the consensus yesterday was that the use of charred material was not a problem. He said that one laboratory intercomparison test had been made, why did we need another? What good could it do to have more than two labs involved? He then gave us an irrelevant lecture on metrology.

Lukasik remarked that we in science tended to argue that more is better (I am sure that is the way suppliers of fighter planes to the US military argue). He thought it was good intellectual discipline to justify every milligram. He concluded that a minimum of three samples were needed. He said he was here as a representative of STURP. STURP would be willing to undertake the responsibility of sampling in connection with all its other tests (I was sure they would be! He was clearly propagandizing for STURP. He was simultaneously trying to carve out a role STURP could play in carbon dating the shroud and making a pitch for STURP being allowed to carry out another series of tests on the shroud.)

Shirley and I had a chance to converse informally with Lukasik again during one of the coffee breaks. He had recently joined STURP after becoming fascinated by the shroud as a result of some article he read in *Reader's Digest*. He said his role was to decide what future STURP tests made scientific sense and what did not. He was one tough, steely eyed individual—an unreconstructed cold warrior in his position of vice president for research at Northrup and clearly another 'true believer' as far as the shroud was concerned.

Chagas, following Lukasik's remarks, said that if the date did not correspond to the crucifixion date, then the whole picture would change. He thought that it was important to concentrate on the problem of dating.

Harbottle, Woelfli, Duplessy, Hall, Otlet and Damon, in turn emphasized that the shroud was not the sort of object one normally encountered in carbon dating. It was special and deserved special treatment. Gonella must know that better than most people.

Woelfli talked about the infamous outlier in the British Museum intercomparison tests and said that it was due to human failure, not in his laboratory but he took the responsibility for it and such a thing can happen again. One of the problems with small samples is that one never knew when the cleaning procedure was sufficient. That was why he thought that anybody who was willing to date the shroud should get a sample.

Hall argued that we were dealing with people outside the scientific community. "Belts, braces and suspenders, this is the approach we ought to take because we are going to be under fire and if we can say that six out of seven agree with each other we are in a much better position than if we say that two out of three agree. I am quite sure that the more we have the more convincing we will be." This comment was interesting because when Gonella finally made the decision that only three labs could participate, one of which was Hall's, he was quite approving on the grounds that the more labs that were involved the greater the chance that one or more would make a mistake and get a wrong answer.

I was then called on to speak. I said I was in essential agreement with my colleagues. I commented on Lukasik's suggestion that the taking of samples be accompanied by an elaborate set of other scientific tests. I noted that STURP's prowess in conducting complex scientific tests was well known and it did them a great deal of credit. In the time it would take for them to get ready, we could already have dated the shroud. I thought we should follow Professor Chagas' suggestion that we settle the question of taking samples for the purpose of dating the shroud.

Gonella said that Harbottle and Hall gave good arguments, political but good, for going to seven samples. This clarified things much better for him. He trembled at Meacham's thought that we must sample as if the shroud were heterogeneous material. He

Paul Damon (Arizona), Willy Woelfli (Zurich), Harry Gove (Rochester) and Shirley Brignall (Rochester)—taken during the workshop.

went on to say that carbon dating had slowed down the other tests rather than vice versa, as Gove claimed. We were apparently going to ask for seven samples for seven labs because of other than scientific reasons—for the reason of greater public acceptance.

Chagas said that STURP was interested in experiments outside the field of carbon dating and this workshop was devoted only to the latter. If STURP wished to somehow be associated with the carbon dating, it was not a decision Chagas could make. He had talked to a number of people about the meeting and they said that any future tests on the shroud depended on what its age turned out to be. The problem was well beyond that of metrology—it must be convincing to the well informed man in the street. Yesterday it was said that we should use uncharred material away from the burned areas if possible and each of the seven labs would get a shroud sample. The Pontifical Academy selected seven labs, so seven labs should be involved. Carbon dating the shroud was the measurement of this century. STURP might want to make further measurements next century. With the exception of Professor Gonella, there seemed to be general agreement that the seven labs should get one shroud and two control samples.

Again the question of making the measurements blind arose. There was considerable sentiment in favour. Finally it was agreed that only six labs would get a shroud sample and the seventh

Doug Donahue (Arizona), Madame Flury-Lemberg (Zurich textile expert), and Jean-Claude Duplessy (Gif-sur-Yvette).

lab would get a dummy sample resembling the shroud as closely as possible. This despite Madame Flury-Lemberg's unequivocal statement that there were no samples that resembled the shroud. The seventh shroud sample would be held in reserve. I was the only person opposed to this arrangement. I privately consoled myself with the knowledge that, at least, I would know whether Rochester received a real or dummy shroud sample merely by weighing the threads.

After lunch, the meeting re-convened. Chagas said we should now discuss the procedures for taking the samples and how they would be transported to the British Museum and then to the laboratories. When the results were obtained, how would they be analysed statistically and how would the final results be made public? He said that Ballestrero wanted a press release after this meeting.

Harbottle considered it essential that the statistical analysis of the data should be made by the British Museum. Each lab should send its raw carbon-14 ratios and their analysis to the museum. The museum would then do the final analysis and arrive at a final age for the shroud following standard carbon dating procedures. It was suggested that the Turin Metrological Institute and someone designated by the Pontifical Academy should also be involved in the final data analysis. I said I thought it important that the labs play a role also, unlike what happened in the interlab comparisons. Harbottle suggested that the lab members get together after today's meeting to design a standard form for reporting their data.

Gonella commented that in addition to the British Museum, the Turin Metrological Institute and the Pontifical Academy, there should be a representative of the bishop present during the data analysis (presumably he meant himself). Ballestrero would not want to be the one to report to the press the results of the carbon-14 measurements on the shroud. The results should be published by the scientists themselves. As a matter of courtesy, the bishop should be the first to know. He said it did not matter whether the analysis were done in Turin or London.

More discussion of the shroud samples followed. It was decided that they should not be cleaned before being handed to the laboratories. Each lab would follow its own cleaning procedures. It was also agreed that if they were unravelled the threads should not be cut into lengths shorter than a centimetre.

As we all should have remembered, the denier of the warp and weft threads of the shroud was known and thus as long as the threads were of sufficient length (e.g. longer than 0.5 cm) the shroud sample could be identified. Blind dating was out of the question. I rather wish now that I had pointed this out when it became clear that it would not be acceptable to cut the threads much shorter than 1 cm. That would have killed once and for all the ridiculous suggestion that the shroud be dated blind and the tedious discussion on how to do it.

Hall then asked if we were clear on the number of shroud samples. Chagas said the consensus seemed to be seven shroud samples and one dummy sample. Six labs would get a shroud sample and one a dummy. The seventh shroud sample would be held in reserve.

I said it was not clear to me who decided which lab would get the dummy. (I had come to the conclusion that if it were up to Gonella, Rochester would not get a shroud sample.) In my view the two institutions that enjoyed the confidence of the public were the British Museum and the Pontifical Academy and I would like the distribution decision left up to them.

The question was raised of who the textile expert would be to oversee the sample taking. Would it be someone provided by STURP? Chagas said that, with all due respect to STURP, Flury-Lemberg was the proper choice.

Adler then described the magnitude of the attack STURP wished to launch on the shroud—there would probably be four or five 'conservation textile people' involved—in addition to many other experts and observers. Chagas said he would like to plead that Flury-Lemberg be involved. She has taken part in all this discussion and could make an important contribution. He said he was not dying but that he was an old man and would like to see this dating done as soon as possible, so when could STURP take on this job? This comment appalled me but I remained silent. As far as I was concerned, STURP would never take on the job.

Dinegar said STURP could remove samples as soon as they had permission to do so. Madame Flury-Lemberg was probably acceptable to STURP. He then said they needed a few hours to decide whether they could indeed accept her. (I thought to myself did Dinegar think he had any credentials to judge her competence as a textile expert—or did she have to satisfy some other arcane STURP criteria?) Dinegar asked, just for the record, whether she were willing. She nodded yes.

Gonella asked Dinegar what lead time was needed by STURP to pack things, organize things, transfer equipment, etc. Dinegar replied that, to just take samples, not much lead time was needed. Chagas asked whether STURP could be ready by March or April? Dinegar said STURP had been hoping to get permission to proceed by then. I could see that Lukasik, by this time, was really licking his lips. He said it should be possible for STURP to carry out its grand plan for tests on the shroud by then and, incidentally, to take samples as part of a broader set of experiments. If permission were received relatively soon, they could be ready by April.

I said that I just wondered whether the STURP group would not want to know, for example, if the shroud was 600 years old? Would that not change their plans concerning the great battery of tests they propose? Surely a 600 year old piece of cloth would be of little interest to them.

At first there was a chorus of 'no's from Dinegar, Lukasik and Adler but they realized that was not an appropriate response. Dinegar said it would be of interest to them to know the age of the shroud. Whether knowing it did not date to the first century would change their plans, he could not, at the moment, say.

Lukasik then launched into a most remarkable speech on how it would really not make much difference if STURP just took samples or if they carried out all the tests they lusted to do. The samples would have to be characterized (whatever that meant), the shroud would have to be taken from its present location, a number of people would have to look at it, to confer thoughtfully, and there would be a lot of overhead costs connected with any sampling. He said the suggestion I raised was a good one—did one want to bring over ten tons of equipment if the shroud were too young? (Too young for whom? I wondered.) He said proper imaging must be done (he was beginning to grope for reasons). Finally, he said that the question I raised was a good one and one they would have to consider carefully.

I remarked that, as far as equipment to do sampling on the shroud was concerned, a pair of scissors appeared to be sufficient. The shroud just needed to be laid out on a table—it did not even need to be completely unrolled. Some of the patched areas and the edges could readily be revealed. I said, like Chagas, I was also growing old and even had the wild notion at one time that we could have taken samples during this meeting. I was clearly too optimistic, to put it mildly. But I would have thought that maybe in a month or even two weeks we could take some samples if we separated the carbon dating from all the other tests.

Chagas said we were here to discuss the problem of carbon dating and we had completely drifted away from it. We should dissociate STURP tests from radiocarbon dating because carbon dating was an urgent thing. It might or might not interfere with STURP's work. To take samples, we did not need an enormous amount of machinery. He said he believed that carbon dating could be done very soon. It could be done as soon as Tite obtained the samples. We did not have to wait for the very important work of STURP.

Meacham said he did not agree with Chagas that we had drifted away from the main objective. He said it was inconceivable to him that the whole proposed test programme should be curtailed. He said an archaeologist would never pack up his work on a dig just because some date came out to be younger than he had expected. The whole huge testing enterprise should go forward. Try as I might, I could not take Meacham seriously. He had such a boyish enthusiasm for archaeology and yet his attempted analogies between that field and carbon dating the shroud seemed far fetched to say the least. He should never have been allowed to play a role in the workshop.

Dinegar seemed now to accept what appeared to be the inevitable—the great STURP assault on the shroud was stymied. He said STURP accepted the responsibility to take samples for carbon dating. I thought to myself—over my dead body they will.

At this point, Harbottle dropped what turned out later to be a real bombshell. He said what many of us must have been thinking, particularly after being exposed to Gonella's slide show. He said he agreed that conservation was very important. As far as he knew, no adequate conservation of the shroud had yet been carried out. The STURP tests conducted so far involved exposing the shroud to powerful visible light, ultraviolet light and X-rays. They could have been very harmful to the shroud. He was surprised that morning at how bright the lights were. Museums protect their valuable cloth from light, especially ultraviolet light. He also could hardly imagine that the collection of surface detritus with sticky tape could be beneficial to the shroud. Textile experts should look at the shroud for questions of conservation; at that time samples could be taken. The STURP tests constituted a stress no matter how careful they might think they were being. Conservators must look at the shroud and decide what tests could be carried out. As far as he was concerned, many of the STURP tests were very intrusive indeed and could have caused image fading.

Lukasik hung in there like barnacles to a ship's bottom. That was just what STURP had in mind, he said. Every move STURP made would be under the control of world renowned textile conservators. Textile conservators would have control at all times. (It is noteworthy that when STURP carried out its assault on the shroud in 1978, conservation was the furthest thing from their minds—it was amazing what a reformation they had undergone!)

Dinegar asked if we were now agreed that STURP and Madame Flury-Lemberg would take samples. Chagas said that at the beginning of May, samples would be taken under the direction of Flury-Lemberg, after conservation assessment.

Chagas then said he assumed that we were all pledged to keep things as secret as possible until the final result could be announced. As true scientists, we must not lie, so if we were asked whether we were dating the shroud we could not deny it. We could say we were but we did not have the results yet. That was probably the maximum secrecy we could maintain.

Chagas said that the cardinal had confided to him that he would not refuse to accept the report. There would be a press release the day after tomorrow. He was sure the cardinal would give the green light to send the report to the Vatican. He then discussed the possibilities for publication of the final results and said it was up to the scientists involved to decide what journal was most appropriate. The second day's meeting adjourned at 5:45 pm.

Representatives of the radiocarbon labs met in the conference room in the seminary immediately following adjournment. It was agreed we would send Tite an analysis of our method of making a carbon dating, an account of how we decided on errors and also how we did the interlab comparison measurements. The best way to do this would be for Tite to send us all a sample to measure. We would then send him back our results and details on exactly how they were measured and the calculations that went into the final answer. It was thought that by the end of October 1987 the AMS measurements could be completed. The small-counter labs take longer and their results would probably not be ready until the end of the year. This assumed the samples would be received in May 1987.

It was during the meeting of the above group that I was called out to talk to Madame Flury-Lemberg. She was very distressed—she had been invited to have dinner with the STURP representatives and said she was under pressure to formally join STURP and sign their secrecy oath. I told her there was absolutely no need for her to do this and advised her strongly not to. She is a very gentle person and I considered it absolutely unacceptable for STURP to pressure her to become a member and to sign any secrecy agreement.

The third and last day of the workshop began at 9 am. Chagas was wearing his wife's scarf (Chagas referred to it as a foulard) as a sling. His hand had become quite swollen over night. He said Cardinal Ballestrero would address the meeting at 2 pm that day. He asked for a report on the meeting the lab people had the previous evening. Hall summarized our discussion. Chagas said he would say that the results would be announced before Easter 1988. He said I had asked for the floor.

I started by saying that I wanted to speak briefly about the question of timing. The timing seemed to be set by the fact that STURP wanted to mount a major enterprise on the shroud. It would involve bringing enormous amounts of equipment to Turin. I had the simple minded notion that for carbon dating all one needed is a pair of scissors in the capable hands of Madame Flury-Lemberg. In order to record the data I saw no need for digital optical analysers or VP-8 image analysers. It seemed to me that a tape measure and a box Brownie would be sufficient. What we were talking about here was taking a sample from the shroud to carbon date—that was what this meeting was supposed to be about. I saw no need for a delay until May.

I continued by noting that this delay, and—I addressed this remark to Professor Gonella—a previous delay in holding this meeting for reasons that were still not clear, were not occasioned by the carbon dating community. The carbon daters have been ready to accept samples for at least a year. Now we faced another delay until an enormous enterprise was organized that I thought could and should be dissociated from the sample taking.

I said I was also surprised that the sample taking was to be in the hands of STURP. There were many of us, speaking very frankly, who were concerned about STURP being the organization to do the sampling. Eventually the samples would be turned over to the organization that most of us in the carbon dating community had great trust and faith in, the British Museum. I wanted to see a simple operation of removing samples with the help of Madame Flury-Lemberg, witnessed by representatives of the carbon dating laboratories. I said I saw no reason at all, and I was being more forthright and less diplomatic than I should be, that our carbon dating enterprise had to be connected in any way with any of the enterprises that STURP wished to carry out. I said I felt so strongly about this that I would have to consider the question of our laboratory's participation if STURP played any role at all.

On the question of whether each lab received a shroud sample, I pointed out that physics was fun, especially when occasionally one got to make a measurement that would have enormous appeal to the public. Normally in our lab we measured such things as the shapes of nuclei—were they round or football shaped? Such research generally caused the public's eyes to glaze over. When it came to the shroud, there was enormous public interest. One of the rationales I had used for our lab being involved in the shroud was that occasionally we should do something that was of interest to the public. After all, it was their tax money ultimately that funded our laboratory. If I went back to my lab and said to the people there we had three samples—it may be that one of them was not the shroud, but after all it was a cooperative effort—I was going to have great difficulty getting them enthused. Whether one called such sentiments political and not scientific, they were very real.

On the question of blindness, if we had three samples and did not know which one was the shroud, we would go ahead with the measurements and whatever our prejudices might be about the shroud's date, they would not affect the measurement. It was true that in making a measurement sometimes one involuntarily got the answer one wanted by somehow unconsciously manipulating the measurement or the apparatus and I assumed this was the reason for making the measurement blind. We did not need to go the additional step of not supplying a shroud sample to one lab.

There were some who argued that making the measurement blind would protect us from the press, especially if we did not know whether we had a shroud sample. We were not babes in the woods—we could handle ourselves with the press. For some eight years now the old timers in STURP had been subjected to blinding publicity from the world press—their heads might be figuratively bloody but, if I looked at Dr Dinegar, his head seemed unbowed. He seemed to have survived the assault of the press. We were all grown-up people and I thought we too could survive. It was something I was not going to worry about.

On the question of breaking secrecy if we knew what the result was, then any lab doing so, to use a British expression, should be sent to Coventry. It would risk being dishonoured in the carbon dating community for all time. So we had a very high motivation to maintain secrecy and to see that members of our respective labs did so as well.

So far I had heard no compelling reason for not supplying a shroud sample to each lab. If that were not done Rochester would withdraw from the dating.

Donahue said it might be coincidental but he too had been thinking about this the last 12 hours. He said, as a scientist, he would like to know what sample he was dealing with. He would like to watch the sample being removed from the shroud, to put it in his pocket, bring it back to his lab and make the measurements. He said that might sound unreasonable but to him that was the essence of doing the experiment. He hoped the group could make arrangements to make him feel better about the whole thing. He said he was also upset at the idea of shredding samples. It could make pre-treatment riskier. He believed all of us were capable of being objective and we did not need a great number of safeguards, especially if these safeguards were detrimental to the overall measurements.

Harbottle said that, along with Donahue, he worried about shredding samples. It would make cleaning more chancy—there would be a greater chance of absorption of impurities. He would want to be present when samples were taken.

Meacham said he would like to start out by supporting Dr Gove on a couple of points—perhaps the only points he could support him on. He had never supported the idea of blind measurements—the labs should come to see the sample taken, give them their own code numbers and take them away with them. Second, if one dropped blindness then there was no reason why each lab should not have a shroud sample.

Otlet said he, of course, would love to be present when the sample was taken but if he could not be he would be quite happy to have Tite act as his representative.

Duplessy said he would want to see the shroud samples being taken, would want to take his sample back to his laboratory, and would want to follow it every step of the way through the AMS dating procedure. He wanted the samples to be left whole—he would have scientific problems with shredded samples. He would like all the labs to be present when the samples were cut and then to receive them and the control samples on the spot.

Hall said he would like to see the samples cut, but then Tite could put them and the controls in envelopes and hand them to the lab representatives. If the samples were left whole it would be easy to identify the shroud. The idea that one could find other samples of the same weave was unlikely. So blindness becomes impossible. He said he thought that mattered, because if we did not make the measurements blind we would be open to criticism. I found Hall's arguments very confusing. He seemed to be saying there was no way in which the measurement could be made blind but one had to do so anyway.

Woelfli said he had checked with the person in his lab who would be doing the pre-treatment and she was horrified to learn that the samples might be shredded into small pieces.

Tite said that clearly there could be no question of blindness unless the samples were broken down to some extent. It was not he who was pushing for blindness. He would like to send the labs a linen sample shredded to the extent that he thought was safe and get their reactions.

Gonella gave another of his tiresome monologues. He was distressed that nobody appeared to trust anyone else. Should he insist that he follow each sample from the ion source through the terminal to the final counter? (At the time this struck me as an inventive way to dispose of Gonella.) As far as dating the shroud was concerned, Turin could have proceeded like any archaeologist in the field who wanted an object dated—just send shroud samples to a couple of labs of their choice (in the event that is exactly how Gonella proceeded). But we had decided to proceed differently.

He criticized the decision to involve seven labs. He said that $12\frac{1}{2}$ cm^2 would be the largest sample ever taken from the shroud. There was no technical justification for this many labs and this much sample. He said he did not understand why the labs had to be present at the sample taking to be able to put the samples in their pockets. Did we not trust the British Museum? Finally he said that Turin had not asked for the dating to be done blind. If it were not to be done blind then there was certainly no need for dummy samples and seven labs.

He ended with the remark that "we all know that Professor Gove doesn't like STURP". He noted that I said that many others felt the same way. He was very interested in this—what was the concern exactly?

Chagas censured him by saying he wanted to avoid personal conflicts.

I apologized to Chagas for interrupting him but I had just been accused of not liking the STURP people. I just wanted to say that this was (a) untrue, and (b) irrelevant.

Chagas then said he wanted to assure us he had confidence in us and we should have confidence in him. His coming here as president of the Pontifical Academy was to work out a plan for carbon dating. The shroud was not an archaeological dig— "you must excuse me Dr Meacham". The responsibility of taking samples for carbon dating was the responsibility of the carbon daters—samples should be taken by Madame Flury-Lemberg. When the carbon-14 work was done, STURP could come in. Now what he was about to say might offend Dr Dinegar but their tests, e.g. irradiating with 12 keV X-rays, could be damaging. They must have had an effect on the image. The obtrusiveness of carbon-14 dating was mild by comparison. The samples would be taken outside the image. If the shroud were younger than 2000 years, interest in it would decline—certainly on the part of the general public. We had fixed a date (i.e. samples early May 1987, announcement Easter 1988) and anyone from the seven labs who wished could come and witness the sample taking.

Chagas then asked Giovanni Riggi to explain what was involved in taking the shroud out from storage. Riggi is a Turin microanalyst with the Italian equivalent of a Bachelor of

Science degree. He is the Director and General Manager of 'Studio Progettazioni Riggi', and he invented an instrument called a 'divaricator' that was used to slip between the shroud and the backing cloth. Whatever his credentials as a shroud expert are, he was chosen to remove the sample from the shroud for carbon dating in 1988.

Riggi spoke in Italian but it was translated. He said the problem of getting at the shroud was not quite as simple as some might think. Much logistics was involved. The shroud was in a vault over the altar in the chapel. High ladders were needed. Permission to do so was required. The condition of the atmosphere was important—dry air was needed. A safe clean location and surgical conditions were required. The whole procedure needed to be repeated to put the cloth back. It was a very fastidious and complex procedure to roll it up again and so on and so forth. (When the pope visited Turin in April 1979, they instantly took it down and displayed it for him. See *The Image on the Shroud* by H David Sox. Riggi made it sound as if it were an operation requiring a major firm of riggers—or possibly the US Corps of Engineers. It is also interesting to note that when Riggi finally cut a sample from the shroud he wore no gloves!—as revealed in the video taping of the operation that he arranged.)

Lukasik spoke next. To me his monologues were almost as sanctimonious, tiresome and tutorial as Gonella's. He said he did not want to complicate this meeting beyond the issue of carbon dating. He said he would like to attempt to allay our concerns. Allegations had been made regarding STURP's intent, so it might be useful to state his role. He had become interested in shrouds (shrouds in general?) several years ago (he did not say, as he told Shirley and me, his interest stemmed from an article he read in *Reader's Digest*). Last October, he had been asked by STURP to coordinate its scientific programme. Authenticity, conservation and image formation were STURP's principle concerns. The STURP proposal was the one that was on the table. It was a start, not an end. He agreed that carbon-14 measurements were the first priority. STURP would never do anything that would harm the shroud. He went on in this sanctimonious way for about six pages in the official transcript.

I asked whether all the labs would get a shroud sample?

Chagas responded by asking what the labs thought about not getting a shroud sample? He said Dr Gove came from a democratic country and would, therefore, abide by the majority.

Harbottle said he did not require a shroud sample. Donahue said he would not be happy if he did not get a shroud sample, but would accept the situation. Duplessy noted he would take more care if he were absolutely sure one of his samples came from the shroud. Hedges said he would do a better job if he knew it was a shroud sample, but if a blind test was needed then he would agree. Otlet and Woelfli agreed with Hedges. Chagas said we would have a blind test. "I do not think Professor Gove was serious when he said Rochester would withdraw." I replied that perhaps Professor Chagas knew me too well—we would not withdraw.

Chagas said that the ruling was that we would have a blind test. One person from each lab could observe the sample taking. Gonella expressed worries about observers from the labs being in the room with the shroud. The room is small. (He also told me later that he was concerned about the possibility that some lab representative might suddenly be seized by an excess of religious fervour and rush forward to lay hands on the shroud. This struck me as so ludicrous I laughed in his face.) It would thus be desirable for the lab observers to view the sample taking by closed circuit TV!

Hall asked what would the press release issued by the Cardinal at the end of this workshop actually say?

Chagas said he had not yet consulted with the cardinal. It should say something like "The cardinal with the Pontifical Academy of Sciences called in specialists in carbon dating and other experts on general problems of the shroud to consider procedures for possible dating of the cloth of the Turin Shroud". It would list the seven labs, possibly also the people who attended, the institutions to be involved in the sampling and data analysis, and, finally, that results should be announced before Easter 1988. (The names of the individual participants did not appear in the cardinal's final press release.)

Chagas ended the morning's meeting by reminding us that we still had to hear from the cardinal and still had a lot of hard work to do in preparing a final protocol. He then announced lunch.

The lunch was sun and wine drenched. Chagas was still wearing a sling as a result of his fall. Shirley and I were invited to sit between Chagas and his wife at the centre of the long table on the balcony that surrounded the central courtyard of the seminary. It was indeed a festive occasion because it now appeared that the way was paved for the Turin Shroud to be carbon dated at long last. And we at Rochester would play an important role in this exciting enterprise. Little did we know what obstacles and disappointments lay ahead!

At 2 pm the cardinal appeared to address us—he shook all our hands. Chagas made a welcoming speech in Italian and then the cardinal addressed us, also in Italian. After his speech Gonella provided an English translation:

He came at the last moment in order to not interfere with the scientific talks that were outside his competence. He wanted to stress the great attention the church pays to serious scientific research. He thanked everybody and, in particular, Professor Chagas. He said he was comforted by the fact that science was not purely aseptic but also responded to mankind's spiritual needs. He hoped that our stay without fog in Turin, which is rare in autumn, was pleasant and that we had talks without fog in this room. He was glad that a climate of friendship had been established and that it would continue. Man needs friendship as much as he needs science. So after wishing us success in our work at the beginning of the workshop he wished us a good trip back home.

The workshop reconvened at 2:45 pm and Chagas presented a summary:

(1) This was the moment for carbon dating.

(2) We would take a minimum amount of cloth to ensure rigorous scientific results and to ensure public credibility but would not include charred material.

(3) For statistical purposes it was decided that 7 labs would be involved, 5 accelerator and 2 small-counter labs.

(4) For logistical reasons, the samples would be removed immediately before the use of the shroud by other groups for other experiments.

(5) Samples would be taken from areas devoid of other possible information content and outside the image by Madame Flury-Lemberg.

(6) Seven samples of the shroud would be taken and 6 shroud plus 1 dummy would go to the 7 labs.

(7) Taking of samples would be witnessed on closed circuit TV by representatives of the labs.

(8) After the measurements were made, the results would be analysed first by three institutions, the British Museum, the Turin Institute of Metrology, and a third person or institution to be

Final words; Carlos Chagas, Cardinal Ballestrero, Peter Rinaldi and Vittorio Canuto.

selected by the Pontifical Academy. A final analysis would be performed involving participation by the laboratories.

(9) By exchange of communications, the radiocarbon groups should devise a common form for reporting the data.

(10) The threads of the samples should not be too short—each lab should use its own method of cleaning.

(11) Two control samples were to be supplied by the British Museum.

(12) A press release would be issued on Friday. This release would state that samples would be taken around 10 May 1987 and the results would probably be made known to the public by April 1988.

(13) Final results would be published as a collaboration.

(14) No laboratory would charge for measurement time.

(15) The certifying institutions (distributors) would be the British Museum (Tite), the Pontifical Academy (Chagas) and the Archbishopric (Gonella).

(16) The analysing institutions would be the Pontifical Academy, the Turin Institute of Metrology, and the British Museum.

After the workshop ended, Canuto, Donahue and I met to draft a text of the workshop agreement to be presented to the cardinal and the pope. Our meeting was held in the main square of Turin at a restaurant table while Donahue and I consumed enormous ice cream sundaes and Canuto sipped Campari accompanied by a plate of olives.

I presented the results of the workshop, including the final agreement, as a poster at the *Fourth International Symposium on Accelerator Mass Spectrometry*, which was held at Niagara-on-the-Lake, Ontario, Canada from 27 to 30 April 1987. It was published as a paper in the conference proceedings that appeared in the journal *Nuclear Instruments and Methods in Physics Research*, Section B, Beam Interactions with Materials and Atoms, volume B 29, Nos 1&2, page 193 (1987).

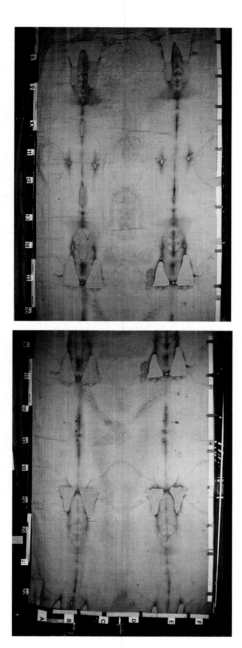

Photographs of the shroud taken in 1978 by a STURP photographer. Top: note the image of a man's head, front and back, in the centre of the photograph. Bottom: note the crossed hands in the upper half of the photograph.

MP Tandem—ion source end.

A detail of the shroud—this piece was removed in 1973 for examination by Professor Gilbert Raes of the University of Ghent's Textile Laboratory. (Courtesy of Vernon Miller.)

1978 Turin crowd scene in the cathedral square, October 1978. (1978 Barrie M Schwortz. All rights reserved.)

A second crowd scene. (1978 Barrie M Schwortz. All rights reserved.)

STURP's cardboard cut-out—created by NASA's VP-8 image analyzer.

Shroud illuminated during the ultraviolet fluorescence photography carried out by STURP in 1978. (Courtesy of Vernon Miller.)

Max Frei (Swiss forensic expert) centre and John Jackson (Captain in the US Air Force and member of STURP) right examining the shroud in October 1978. (Note Jackson's wooden cross.)

Bishop John A T Robinson (Church of England) being interviewed during the October 1978 shroud congress.

Christina Leczinsky-Nylander—producer of Science and Cultural Programmes for the Swedish Broadcasting Corporation—interviewing Michael Thomas 'Religion Correspondent' of Rolling Stone Magazine.

Harry Gove meeting Pope John Paul II—September 1986. (Credit the Vatican.)

Final lunch at the Turin workshop. Carlos Chagas (Pontifical Academy of Sciences), Shirley Brignall (Rochester) and Harry Gove (Rochester).

Meeting in November 1986 at the Algonquin Hotel in New York City with David Sox and a representative of the BBC regarding televising the shroud dating. Sox centre and Gove right.

The array of instruments available for taking a shroud sample for the three laboratories to date. This frame was taken from the BBC's Timewatch programme on dating the shroud. (From the BBC's Timewatch programme. Courtesy of Giovanni Riggi di Numana.)

Scissors in the hands of Giovanni Riggi, a Turin microanalyst. His talk at the Turin workshop stressed the absolutely immaculate conditions under which the shroud must be handled! (From the BBC's Timewatch programme. Courtesy of Giovanni Riggi di Numana.)

Those present at the Arizona AMS carbon dating facility at 9:50 am on 6 May 1988 when the age of the shroud was determined. They include Doug Donahue (third from the left, standing), Tim Jull (fourth from the left, standing), Harry Gove (sixth from the left, standing) and Paul Damon (seventh from the left, standing).

Chapter 7

Dating Delay

Shortly after our return from Turin, the 5 October 1986 edition of *La Stampa*, the Turin newspaper, arrived at the university. It stated that the pope had given his approval for the shroud to be dated by seven laboratories following the Turin workshop agreement. It quoted Luigi Gonella, 'the coordinator of the research', as saying that, because of uncertainties in the measurements, a date between 100 BC and 400 AD might be found. That would be good enough, however, to quiet those who state that the image is the work of a forger. This left no doubt as to what the Cardinal of Turin's science advisor believed the shroud's age to be.

When I discussed this news story with Canuto, he said the pope had not approved the dating of the shroud. The Turin Protocol was typed in Rome and copies would be sent to everyone including the pope and Ballestrero. Chagas was having second thoughts about the tests STURP planned to carry out after the removal of the shroud material. He was concerned about the floodlights and the X-ray and ultraviolet irradiation that would bathe the shroud. In his view, the carbon-14 tests were not only the least stressful and most unobtrusive measurements, but also the most important. Chagas would like advice from the laboratories on the possible danger that these tests might pose to the shroud. Canuto offered to call Harbottle and Donahue and suggested I should contact the other four laboratories to ask them whether they would be willing to give Chagas the requested advice.

The next day, Hall told me that he would write to Chagas thanking him for the splendid workshop and would express his concerns about the STURP tests. He was less worried about the X-rays than the ultraviolet light. One hangs washing out on the

177

line to get bleached in the sun and that bleaching is caused mostly by the ultraviolet radiation.

Woelfli said he would write to Chagas. Later that same day, I contacted Duplessy. He said that he was worried about the presence of the STURP people but did not want to offend them. He had no information on what kind of measurements they want to make and so would not volunteer to advise Chagas. Otlet also declined to offer Chagas advice on the STURP tests.

Harbottle and Donahue both told Canuto they would write to Chagas. Harbottle was quite upset with STURP, especially with Lukasik and Dinegar. He said that there was nothing new in the proposed STURP tests. Canuto remarked that the tables were really turned. Gonella, despite his professed concerns about shroud conservation, was willing to risk its being bleached in the hot glare of STURP's ultraviolet radiation.

In my letter to Chagas dated 7 October 1986, I expressed my admiration for the skillful, diplomatic and firm way in which he had chaired the Turin workshop. I was not just being complimentary—he really had done a splendid job. I agreed with him that carbon dating the shroud was the least intrusive and stressful of any of the hierarchy of measurements so far proposed. If the cloth dated from the first century AD, there should be no further measurements permitted until they had been approved by some reputable and dispassionate international group of scientists.

It became apparent sometime in November or December 1986, that these concerns expressed to Chagas were taken seriously. STURP has never been given permission to go ahead with their measurements.

In mid October 1986, Sox wrote to me asking if I could come to New York on the 6th of November to talk to Neil Cameron of the BBC *Timewatch* programme. Cameron was doing a programme for this series on Sox' book *The Forger* and was interested in doing one on the shroud dating.

On 5th November, just before going to New York, I talked to Canuto again. Chagas and Rovasenda had had lunch with the pope the week of 20 October. The pope had expressed pleasure at how the workshop had gone but had expressed concern about STURP. Canuto said that Chagas had received letters from me, Donahue, Hall and Harbottle concerning the further tests on the shroud proposed by STURP. He had summarized them in a letter for Chagas to deliver to the pope this week. The letter noted that if its age was the historic one, interest in the STURP tests would diminish. If it was 2000 years old, the STURP tests must be scrutinized as carefully as the radiocarbon tests had been.

On 5 November Shirley and I flew to New York to meet with Sox and Cameron. We had dinner with them at the Algonquin Hotel that evening. Cameron, a 32 year old, low-keyed person, was someone to whom I took an immediate liking. He had a 1978 PhD in biochemistry from Imperial College, London. He expressed an interest in doing a programme in this *Timewatch* series on the carbon dating of the shroud and suggested he send a crew to Rochester when we got our sample and started the dating run.

Back in Rochester, Shirley spotted a piece on the shroud in a publication called *Picked Up For You This Week* mentioning the carbon dating of the shroud to be done by seven laboratories. It made reference to an article that had appeared in a journal called *Le Paranormale* by Henri Broch. Shirley had requested the article on interlibrary loan.

The journal arrived and was in French. As part of the article, Broch remarked on the membership of STURP. STURP counts only one agnostic among its forty members. He asked in the article, ironically, what were the chances in choosing at random forty scientists among thousands who were competent in the US to make studies on the shroud, and to come up with a group which comprises thirty-nine believers and only one agnostic? He concluded that STURP was not an organization that would be disposed to have a sceptical attitude regarding the question of the shroud's origin. From everything I knew about STURP, Broch's conclusions were right on the money. It was something that had worried me a lot.

On 17 December I asked Canuto if there were any new developments. He said the letter from Chagas to the pope had also been sent to Ballestrero. Gonella had been shown the letter and had seemed quite composed about it. One would have expected him to be enraged by the thought that STURP might be slighted. Canuto was suspicious of his calm reaction.

I mentioned that I had been contacted by Neil Cameron, a producer of the BBC programme *Timewatch*, about doing an hour long programme on dating the shroud. Canuto said I should write to Chagas presenting my views on having BBC involvement and also noting that the *National Geographic* had been given sole rights to cover STURP's 1978 tests on the shroud. A week later, I did so saying that if he were willing to discuss the matter with Cameron,

I would tell him.

On 5 January 1987, Meacham wrote a letter to Adler, Damon, Donahue, Hedges, Lukasik, Otlet and Woelfli, but not to me. However, someone voluntarily sent me a copy. Meacham said that he had received very disturbing news from Gonella that caused him to write to Chagas.

The letter to Chagas with copies to Ballestrero and Tite stated that he was most distressed to learn recently that certain participants in the Turin workshop had expressed concerns about other planned testing of the shroud and the deleterious impact such tests might have on the relic. These tests were totally outside their area of expertise and questioning them reflected an attitude hostile to the proper scientific study of the relic. This was reinforced by the provocative and improper remark made at the meeting, regrettably allowed to pass without reprimand by the chair, that STURP was on a crusade to study the shroud in the full glare of publicity. (That was probably his version of some quote from me.)

He felt the consensus arrived at in Turin had several major weaknesses. In spite of its shortcomings, however, the Turin Protocol was a consensus elaborated through negotiations. The conclusions were agreed upon in a 'gentlemanly' (had Meacham forgotten there were women present at the workshop?) fashion in a spirit of amity and, more importantly, in a way that gave due consideration and weight to the various interests represented. The timing and relevance of the other testing programmes were discussed in Turin at length and the agreement on them was expressed in Item 4 of the conclusions paper (Item 4 stated that samples would be taken just prior to other measurements). Any attempt now to alter this agreement through back-room dealings appeared to him to be subversive and underhanded. He would not hesitate to withdraw his support of the Turin Protocol (as if it mattered a whit whether he supported it or not) if this very weak provision for the securing of additional contextual information relevant to the samples to be used for carbon dating was not respected. He hoped that Chagas, as chairman and co-organizer of the Turin workshop, would endeavour to ensure that the consensus expressed in the protocol would be strictly adhered to. No faction should be allowed to manoeuvre, through whatever connections they might have, their priorities to the fore or erect obstacles to the procedural steps recommended after having failed to obtain majority support for their position at the workshop.

Either there was a basic agreement on the protocol or there was not. In the latter case, the workshop would have been little more than an academic debate and the authorities responsible for the shroud would be justified in reconsidering the entire programme for radiocarbon dating. He concluded, "With every hope that the C-14 and other testing of the Shroud will proceed smoothly, Sincerely, William Meacham, Hong Kong."

Meacham did not realize it was Chagas who had reservations about STURP's proposed measurements as a result of Gonella's slide show at the workshop. The inputs Chagas had requested from the carbon dating labs merely reinforced his concerns.

Canuto returned from Rome in early February 1987. He said the people in Turin were still dragging their feet. Ballestrero had visited the Vatican recently, but did not deign to visit the Pontifical Academy of Sciences although he was invited to do so. Chagas had dinner with the pope while Ballestrero was in Rome. Chagas reiterated his concerns about the STURP tests and the pope said that he was meeting with Ballestrero the next day. He would remind him that some action must be taken with respect to carbon dating. Canuto felt that the STURP tests were the real obstacle. Gonella must be pushing for them. Chagas opposes them. Canuto had suggested that Chagas send Rovasenda to Turin to meet with Ballestrero on the basis of one priest to another. They were old friends and it could be held without Gonella being present. Chagas agreed to do so.

Canuto said the pope told Chagas that cardinals are powerful. I was beginning to have little doubt about that and to realize that on many matters relating to his brother cardinals, the pope was just another one of them. Ballestrero and Gonella were the key players in this shroud drama. Canuto remarked, with tongue in cheek, "How is the pope to know that Gonella is a second rate scientist and that the rest of us are super scientists?"

On 8 March 1987, Sox called me from London. He said that in his recent trip to Italy he had some indication that all was not smooth sailing with the shroud dating project. He had no doubts that the problem lay with STURP. He had just learned that Don Coero-Borga, who was head of the International Sindonology Centre in Turin, had died this past December 1986. Borga had always been opposed to dating the shroud so we had one less adversary. Plenty remained.

In mid-March, Canuto reported that he had called Chagas. He was laid up with water on the knee from his Turin fall. There was no word yet on dating the shroud. A day or two later, I wrote to Chagas, on the letterhead of the *Fourth International Symposium on Accelerator Mass Spectrometry* that was to be held 27–30 April 1987, as follows:

"I am sure that you are doing everything in your power to move the appropriate authorities in Rome and Turin to action on a decision concerning a radiocarbon measurement of the Shroud of Turin. In view of all the effort you have devoted to this enterprise you must feel as keenly disappointed as I do that so far no further progress has been made. It will not make an iota of difference to me if the Vatican refused to permit the shroud to be dated. A decision, however, should be made soon, one way or the other."

I then informed him that I intended to present the protocol to be followed in dating the shroud as a poster at the *Fourth International Symposium on Accelerator Mass Spectrometry* in April. If I were asked why the dating has been delayed I would say that it was no longer possible for me to conclude that the Vatican or the Turin ecclesiastic authorities regarded the new carbon dating techniques as still unproven. It must be because STURP was insisting that its battery of high technology, invasive and possibly deleterious measurements must immediately follow the removal of a small sample for carbon dating. If the shroud's age was 2000 years, no further tests should be allowed without the same detailed scientific scrutiny that was to be applied to the radiocarbon measurements. STURP's desire to conduct further measurements on the shroud seemed to have rendered them bereft of reason. Maybe the simplest solution was for the Vatican to rule that all further scientific tests on the shroud were in abeyance until adequate steps were taken to ensure its preservation. That would at least provide some logical reason for the delay.

I received a letter dated 19 March 1987, from Mark Plummer, the executive director of the *Skeptical Inquirer*, a journal that investigated claims of the paranormal, asking whether it would be possible for scientific observers from his committee to be present at every stage of the tests. He enclosed a document from the Holy Shroud Guild that stated the project would involve 'no small expense'. He would appreciate it if I could inform them if the testing at my laboratory is being directly or indirectly funded by the Holy Shroud Guild.

In the enclosure from Plummer, there was a letter from Father Rinaldi to 'Friends of the Holy Shroud Guild'. In it he wrote that there was exciting news about the shroud. Within the next few months the shroud would undergo the most thorough scientific examination ever attempted, including the crucial carbon-14 test which would date the cloth with a fair degree of accuracy. He stated that it was the Holy Father's wish that everything be done to solve the mystery of the shroud. He went on to say that it was a very important project. It would involve many qualified experts at no small expense. He said he was counting on the prayers of the recipients of this letter and on any small contribution towards the success of the project they might be able to make. He enclosed a form for people to send to him reading: "Dear Father Rinaldi: Do keep me posted on all shroud developments, I am pleased my contribution will help the scientific research." This was followed by the request to "Kindly make your check out to the Holy Shroud Guild and enclose it herewith".

On 28 March I informed Chagas of the Rinaldi letter. He was amazed. Except for the carbon dating, it was STURP's plans Rinaldi was heralding and clearly Chagas had not been kept informed of anything STURP was proposing. He asked me to send him a cable containing the contents of Rinaldi's letter.

On 10 April I phoned Gonella in Turin. I asked him if there were any new developments on carbon dating the shroud. He replied that Turin was still awaiting word from Rome. I replied that the information I had was that Rome was awaiting word from Turin. He said that his information was different. I then read him the first paragraph of Rinaldi's letter to members of the Holy Shroud Guild. I said the only way I could interpret it was that STURP had big things in the works and I was worried about the impact this might have on the completely separate task of carbon dating. He reiterated that Turin was still waiting for a go ahead from Rome. I suggested that Cardinal Ballestrero might phone the pope and settle the question of who was waiting for whom. Gonella reacted positively to what was, to him, a novel idea and said he would pass my suggestion on to the cardinal when he met with him next Wednesday.

On 30 March I replied to Plummer and said that our lab was not receiving any funding from any religious group. I added that there were many people and organizations who would like to be present at every stage of the test but that would be impossible. I assured him that the radiocarbon measurements would be carried out in a manner that would satisfy even the most sceptical inquirer.

On 15 April I sent a progress report to members of the six laboratories and to Tite at the British Museum. It had now been over six months since the Turin workshop had been held and in which we had agreed on procedures. These had been approved by everyone who had participated, including the representative of the archbishop of Turin, Professor Gonella. I outlined the events that had transpired from the *La Stampa* article of 5 October 1986, to my phone call above to Gonella.

On 10 April Chagas wrote me one of his succinct letters. He said "Dear Gove: I have been in and out on the phone but I think that something has happened which will find a solution for the work of dating carbon-14. But as soon as I have more detailed news I will phone you."

In a letter postmarked 15 April 1987 Sox sent me a clipping from the London *Times* dated 15 April 1987, that was titled 'Science and the Shroud'. It was based on an interview with Professor Edward Hall of Oxford University. Hall asked the reporter to imagine a path about 130 yards wide made of a single layer of sand stretching from the Earth to the Moon. Our task, he explained, was the equivalent of finding a grain of sand in that path that differed slightly from the other grains. That was a measure of the problem being undertaken by seven laboratories in Europe and America sometime in the coming months. Everyone concerned was being deliberately secretive about exactly when (they were being secretive because they had no clue as to when). Further on the article stated that, persuaded that men like Professor Edward Hall (there was a picture of Teddy) had the right technology, Rome had released the shroud for dating. It contained a description of the machine (although it did not mention that it was a Tandetron designed by Ken Purser) that would be used by Oxford, as one of the laboratories, to carbon date the shroud.

What I found most amusing was the statement in this article that Hall was among those who first set out the theory for the machine. Hall admitted that the credit for producing the first machines which actually made the theories work goes to laboratories in the US and Canada. He said his was built in 1978 and since then it has conducted several thousand archaeological experiments. (Hall, of course, made no contribution whatsoever to either the theory or development of accelerator mass spectrometry.)

Hall went on to say that he would not know the results of the test until they were announced by the Vatican because he would have no way of knowing which piece of material was from the shroud. As an agnostic, he did not believe in any supernatural explanation for the shroud's images. He added that he viewed Christ as a historical individual with a powerful personality. He admitted that it is possible that, in some way we do not currently fully understand, some kind of impression from him was transferred to the shroud. However, if the carbon date turned out to be around the start of the first century AD, he would then find it difficult to dismiss the shroud's authenticity.

On 25 April Sox sent me a letter he had received some time earlier from Elizabeth Crowfoot. He noted that Elizabeth and her mother were leading experts on ancient textiles. In the letter she had commented on the question of the herringbone weave of the shroud. She said such a weave was not at all common 2000 years ago, particularly in linen. That weave itself was a good indication that the shroud did not date back 2000 years.

The issue of *Science News* dated 25 April 1987 described some of the events that had occurred at the spring national meeting of the American Chemical Society in Denver, Colorado. It included an article titled 'Carbon-14 Dating for the Shroud of Turin?' It claimed that carbon-14 dating and a battery of other tests to help scientists gauge the shroud's age, place of origin and, inferentially, its authenticity, could begin as early as this year. The information came from Garman Harbottle, senior chemist with the Brookhaven National Laboratory in Upton, New York and a co-coordinator of the Shroud of Turin Research Project, STURP. It mentioned the three day meeting that had been held in Turin to determine the procedures for dating the shroud—there would be seven laboratories and the carbon-14 measurements would be done blind.

It went on to say that in one test to help determine the shroud's origin, researchers would analyse the isotopic ratios of heavy oxygen to regular oxygen and of deuterium (or heavy hydrogen) to regular hydrogen. It quoted Robert Dinegar as explaining that these ratios might help establish whether the flax used to make the shroud's cloth originated in a climate that was warm or cold, dry or moist. Those did not seem to me to be particularly urgent or even relevant questions to answer. They were, however, quite typical of STURP's ideas for scientific tests on the shroud.

Both Dinegar and Harbottle were confident the tests would occur. Harbottle, who was eager to pick up samples, said he was talking to a travel agent and that these tests would happen.

The *Fourth International Symposium on Accelerator Mass Spectrometry* was held between the 27th and 30th of April 1987 at

Niagara-on-the-Lake, Ontario, Canada. It was organized jointly by the University of Rochester, the University of Toronto and Mc-Master University. The meeting was the tenth anniversary of the first measurement of carbon-14 in natural material by tandem accelerators. One of the sessions included the history of the development of accelerator mass spectrometry. The after dinner talk at the symposium banquet was titled 'The Shroud of Turin: Relic or Icon' and was given by W S A Dale, chair of the Department of Visual Arts at the University of Western Ontario in Canada. He was a friend of Ted Litherland's who had told me about his interest in and studies of the Turin Shroud. Dale's talk was published in the symposium proceedings and, in its conclusion, Dale stated that the shroud's most probable date would be somewhere between 1000 and 1050 AD. The conference summary was given by a prominent nuclear physicist from the University of Witwatersrand in South Africa, J P F Sellschop. He noted that, "The Shroud of Turin continues to fascinate: serious discussions were reported with relevant curators, and a quite outstanding art historical address was presented on the seminal question as to whether it is a relic or an icon—more than one scientist was heard to hope that the solution would be long delayed for the sheer pleasure with which it continues to tantalize us."

At the conference, I presented a poster on the conclusions and the procedural steps which were agreed to at the Turin workshop. It was published as a paper in the proceedings of the meeting in the journal *Nuclear Instruments and Methods in Physics Research*.

I organized a meeting of those people at the conference representing the five AMS laboratories that had participated in the Turin workshop to discuss the inordinate delay in a decision to proceed with the carbon dating of the shroud. I was asked to write to Chagas re-affirming our support for the protocol we had all agreed to at the workshop and to press for action.

The next crisis occurred when I got back to Rochester. An article had appeared in the 27 April issue of *La Stampa*. The headline read 'You Shall Know the Age of the Shroud' and under that: 'The carbon-14 tests will be done'. The opening paragraph of this article stated that in 1987 the age of the shroud would be known. Samples will be extracted before the end of the summer. The measurements would be made in two or three laboratories by two research methods after permission from Cardinal Anastasio Ballestrero was forthcoming. This announcement had been made the previous day by Professor Luigi Gonella of the Turin Polytechnic.

It went on to say that it might be possible to know the results by the end of the year, or at the latest in the spring of 1988. Gonella was quoted as saying that soon after that there could be new developments, extremely complicated scientific instrumentation will come back to scrutinize the mystery of the likeness of a man after the passage of thousands of years. (Again Gonella confirmed his deep-seated prejudices concerning the shroud's origin.)

There was one interesting and at least two troublesome implications in this article. It was interesting that two methods would be used. That meant both the AMS and the small-counter decay counting methods would be employed. One troublesome statement was that the number of laboratories would be reduced from the original seven to two or three and the other that the carbon-14 tests would be just one of a whole vast panoply of tests presumably carried out by STURP. Both caused me great concern and I decided that we would have to try to do something about this as quickly as possible.

I composed a telegram to be sent to the senior representatives of the six other radiocarbon laboratories and to the British Museum which went as follows: "I propose to hand-deliver the following letter to Professor Chagas when he is in New York on May 16 or 17, so I would appreciate a quick response. In my view, Gonella and STURP are being deliberately mischievous concerning carbon dating. If the Turin workshop agreement is not followed to the letter, I am no longer willing to be involved. Please approve this letter." The text of the letter follows:

"Dear Professor Chagas: A meeting was held at the Pillar and Post Inn in Niagara-on-the-Lake, Canada the site of the 4th International Symposium on Accelerator Mass Spectrometry on Thursday, 30 April 1987 concerning radiocarbon dating of the Turin Shroud. Present were representatives of the 5 AMS laboratories who will be involved in the measurements, all of whom with the exception of the representative of Oxford were present at the Turin workshop. Since this international meeting concerned accelerator mass spectrometry, AMS, there were no delegates present from the 2 counter laboratories at Harwell and Brookhaven. As a result of the meeting, the undersigned wished to reaffirm their strong, continuing support for the conclusions and procedural steps agreed to by the delegates to the Turin workshop of September 29 to 1 October and in particular: (a) all seven laboratories must be involved in the tests; (b) Madame Flury-Lemberg of the Abegg-Stiftung must be responsible for the selection and actual removal of the material from the shroud; (c) representatives of all seven laboratories should be present at the actual sample removal; (d) a representative of the Pontifical Academy of Sciences, the British Museum and the Archbishopric of Turin will supervise the shroud samples from the time of removal to the time of their delivery, also with a dummy sample and control samples to a representative of each of the seven laboratories. We emphasize the above because of a report in the 27 April 1987 issue of *La Stampa*, the Turin newspaper, attributed to Professor Luigi Gonella, that the carbon-14 measurements will be carried out in two or three laboratories.   That so directly contravenes the Turin workshop agreement that it could severely jeopardize the carbon dating enterprise. The people present at the Niagara-on-the-Lake meeting were S L Brignall, Rochester, C R Bronk, Oxford, P E Damon, Arizona, D J Donahue, Arizona, J C Duplessy, Gif-sur-Yvette, H E Gove, Rochester and W Woelfli, ETH Zurich."

Only Bronk had not attended the Turin workshop. I had asked the senior people of the seven laboratories and the British Museum to please respond by telephone their agreement or otherwise, that this message be delivered to Professor Chagas.

The responses that I got from the various labs to this telegram were particularly interesting in view of subsequent events.

The telegram was sent on 6 May 1987. On 7 May Jean-Claude Duplessy phoned me from Gif-sur-Yvette. He entirely agreed with the contents of the proposed letter to Chagas, he was concerned about complications that may be caused by STURP and repeated that he had serious reservations about them.

Later that same day, Hall phoned from Oxford. He said that he was worried that the letter might complicate matters. He thought Chagas favoured only two or three laboratories being involved. I said it was my impression that Turin was by-passing Chagas and that this letter might strengthen his hand. Chagas had never expressed a preference for any less than the agreed upon number of seven laboratories. Hall thought that if a decision were made to reduce the number it would mean starting again. He agreed to sign the letter. It seemed to me Hall clearly opposed the idea of a reduction in the number of laboratories.

He later changed his mind.

About an hour later, Donahue phoned from Arizona. He said he was in complete agreement with the letter and so was Damon.

The next day, 8 May, I got a bitnet message from Woelfli in Zurich in which he stated that he fully agreed with all the points made in my letter to Professor Chagas and that he would be glad to sign it. He stated categorically that he was not willing to be involved if the Turin workshop agreement were not followed to the letter. He also changed his mind later on.

On 11 May I sent the letter to Professor Chagas. It was signed by the heads of the five AMS laboratories who had given me permission to submit it. On the same date, I sent a second letter to Chagas enclosing a copy of a letter Cameron had sent him and which Cameron sent by fax to me. In my cover letter I again endorsed the involvement of the British Broadcasting Corporation in the shroud sample taking operation.

The 7 May 1987 issue of *Nature* published a letter by Denis Dutton from the School of Fine Arts of the University of Canterbury in Christ Church, New Zealand. He expressed concerns about the protocol for dating the shroud as he understood it. He wrote that repeated inquiries to Tite and others on this matter had so far elicited no satisfactory answers. He went on to say that after years of discussion there was agreement to go forward with carbon-14 tests on the Turin Shroud, but so far without due regard for an open disclosure of the procedures for taking the samples. He said there must be no hint that, for example, fibres of mummy linen might have been supplied to the laboratories rather than the actual shroud samples.

Tite replied in a letter to *Nature* of 11 June 1987. He assured Dutton that all the institutions involved were fully aware of the crucial need to insure that the chain of evidence remain unbroken. It was to meet this need that the British Museum accepted the invitation to act as guarantor and independent observer. He ended by noting that these procedural steps have yet to be finally agreed by the Pontifical Academy of Sciences and the Archbishopric of Turin, so he was not at liberty to divulge their details.

I also replied to Dutton's letter (this was published in *Nature*'s 25 June edition). Before submitting this letter, I obtained approval from Professor Chagas at a meeting that Shirley Brignall and I had with him, to be described below. In my letter I pointed out that the conclusions and procedural steps agreed to at this workshop had been presented as a poster at our accelerator mass spectrometry conference at Niagara-on-the-Lake 27–30 April 1987 and that they would be published in the conference proceedings. They were not confidential. I ended by saying that it was clearly important that the most significant scientific test on the Shroud of Turin, namely radiocarbon dating of the cloth, should be carried out in a manner that would convince people like Dutton that the results, whatever they might be, were believable.

On 10 May an article appeared in the *New York Times* concerning the Vinland Map. A friend of mine at the University of California at Davis, Tom Cahill, Director of the Crocker Nuclear Laboratory at that institution, had employed the cyclotron in his laboratory to analyse the trace element composition of the map using a non-destructive technique called PIXE (particle induced X-ray emission). He concluded that the ink on the map contained only extremely small traces of titanium, amounts that were quite consistent with it being a genuine medieval document. In 1974, Walter McCrone had concluded that the white ink on the map had been made from the pigment anatase (titanium oxide). Such white pigment was invented in 1917 and to Walter this proved the map a fraud. McCrone hotly contested Cahill's findings and fired off an angry letter to him stating, in effect, that war was declared. In an interview the *New York Times* conducted with him, McCrone also branded as fraudulent the Shroud of Turin. He was clearly trying to re-burnish his image as the world's leading iconoclast. As far as the Vinland Map is concerned, I would put my money on Cahill and PIXE. Carbon dating will decide the question of the shroud. It will also settle the question of the Vinland Map. The problem McCrone has is that his scientific techniques are unsophisticated compared to AMS and PIXE.

On 12 May I contacted Chagas. He told me he would be in Boston from the 18th to the 24th of May and was leaving on the 25th for Rome. I should phone him on 19 May at the Copley Plaza Hotel in Boston and could meet him that night. Chagas was going to Boston for a medical checkup.

On 19 May Shirley and I flew to Boston. We met with Chagas at his hotel suite at the Copley Plaza Hotel around 9:20 in the evening. Later we were joined by Mrs Chagas. Chagas described his medical woes, mainly as a result of his fall in Turin. He was in Boston to go to the New England Hospital for extensive tests that would start the next morning at 9:30. Chagas planned to leave for Rome as soon as possible and would be there until about 20 June. He said that he had sent two letters to Ballestrero with respect to the shroud dating but had received no reply. Rovasenda, his deputy in Rome at the Pontifical Academy, visited Turin, met with Ballestrero, but no decision on shroud dating resulted.

Chagas mentioned what he described as an insulting letter he had received from Meacham. He did not specify the nature of this letter but I immediately knew it must be the one I had received a copy of. He said that the tone of Meacham's letter was such that he was unable to answer it. He asked Dardozzi, who would be Rovasenda's replacement, to do so. Dardozzi did and Meacham sent Chagas a letter of apology. Chagas also told us that a transcript of the Turin workshop proceedings was available and I could write to Dardozzi for a copy, which I did when I got back to Rochester. In a phone call to Canuto on 20 May I reported on my visit to Chagas. Canuto said he had drafted the reply to Meacham that Dardozzi signed. He said that he had known about this letter some time ago but had not wanted to tell me about it. Maybe he thought it would give me an even worse impression of Meacham than I already had.

When I questioned Chagas about the delays in shroud dating he said there was a nine-month gestation period for the College of Cardinals. They worked on that time scale for all their operations. At the age of 80, Cardinals give up administrative responsibilities, which is why Ballestrero would be stepping down as archbishop of Turin in a short time. Chagas said that the pope had given Ballestrero the complete right to make decisions regarding the shroud tests. Chagas read my proposed letter to *Nature* in reply to Dutton and approved it. He said that Dutton had never contacted him.

I left Chagas all the relevant documents dating from December 1986 to May 1987. He seemed in very good spirits and his health seemed fine. He and his wife were as charming as ever and it was a very helpful and cordial meeting.

Another month went by and there was still no decision on dating the shroud. Clearly it was due to STURP's implacable desire to be part of the process. I decided to write a firm letter to Professor Chagas, which I sent to Canuto in mid-June. He had agreed to deliver it when he met Chagas in Rome in a couple of days.

In the letter, I sharply criticized the way in which the Turin authorities had handled matters concerning the shroud in the past and the way they were continuing to mishandle them. I noted that the shroud had been subjected to a number of scientific tests of dubious value carried out in ill conceived ways by scientists of unknown reputation.

First was the investigation by Professor Gilbert Raes of shroud samples removed in 1973. Not only were his discoveries of minimal significance, but his care of the samples and their subsequent control by Turin was so careless that the samples were judged not to be suitable for carbon dating.

This was followed in 1978 by the STURP tests carried out by people who were already convinced they were dealing with Christ's shroud. Not only did these measurements yield negligibly significant results, but they subjected the shroud to a number of intrusive stresses.

I stated that almost every aspect of the STURP organization was distasteful to many other scientists. This included their clear religious zeal, their questionable sources of support, their military mind set and, last but not least, their assumption that the Turin Shroud was their property as self-appointed investigators of its origins and properties.

Now, however, the Pontifical Academy of Sciences, under Chagas, had a chance to change all this. I went on to make some comments and suggestions noting that, in so doing, Chagas might consider me presumptuous. Among these were the following:

1. Without the continued leadership of Chagas as president of the Pontifical Academy, the present carbon dating consortium would probably become disenchanted and withdraw their participation. This action would be guaranteed if STURP participates in the carbon dating enterprise in any way.

2. If the delays in carrying out the carbon dating were sudden concerns for conservation, then conservation experts should be contacted by the Pontifical Academy and not by STURP.

3. At the American Chemical Society meeting in Denver, it was stated that shroud samples would be removed from behind the patches for stable isotope ratio measurements by STURP. Such measurements would tell whether the flax from which the shroud linen was made was grown in a climate that was warm or cold, wet or dry! I described this, quite charitably, as outrageous nonsense and asked whether there was nothing that could be done to hold STURP in check.

4. I also reminded Chagas of the pressure STURP representatives had subjected Madame Flury-Lemberg to at the time of the Turin workshop and, exaggeratedly, compared it to the Spanish

Inquisition. I noted that Madame Flury-Lemberg was a gentle person quite unschooled in dealing with people like Mr Lukasik and his colleagues like the Reverend Dr Dinegar. I observed that the STURP members had been extremely discourteous to Madame Flury-Lemberg and had pressured her unduly.

I had hoped the involvement of the Pontifical Academy of Sciences under Chagas' leadership would bring a proper degree of international scientific dispassion and integrity to the scientific endeavours to solve the mystery of the Turin Shroud. So far it had not because, clearly, he was unable to control the antics of STURP. One would be amused by the whole farce if one did not feel so saddened by the consequences STURP's activities would have in elucidating the mystery of the most important relic or icon—whichever it turned out to be—in the Christian world.

"In conclusion, although I am sure all of us who will be directly involved in the carbon dating hope the shroud will be subjected only to sensible and prudent scientific examination and testing, there is nothing we can do about whatever activities are being planned by STURP. What is in our power, however, is to ensure that STURP plays no role in carbon dating. STURP is nowhere mentioned in the Turin Workshop Protocol. We count on you to ensure that at least this one measurement will be carried out in a credible way without being tainted by STURP. After a sample is removed for carbon dating STURP can carry out any measurement they wish, however frivolous, as far as I am concerned."

It was a tough letter—not couched in diplomatic niceties—but it reflected the frustration and annoyance I felt at the way things were going. I never received any reply from Professor Chagas nor any indication of how he reacted to it. By then he may have realized, as I did not at the time, that his boss, the pope, had cut him out of the action.

On 29 June 1987, I phoned Canuto to find out whether he had an update with respect to carbon dating. It turned out to be an extraordinarily timely phone call. He told me Chagas had received a copy of a letter addressed to Ballestrero from the pope, but executed by the Cardinal Secretary of State, dated 21 May 1987. It contained instructions on the procedures to be followed in radiocarbon dating the Turin Shroud. Ballestrero had now had this letter for over a month.

Canuto indicated that, because the letter was from the Secretary of State to the Cardinal of Turin with only a copy to Chagas, he was unable to divulge its contents. Because the procedures differed substantially from the Turin Workshop Protocol, the question was how we could use the key information contained in it, even though we were not to know in detail what it was. We agreed that Canuto should tell the US laboratory people what he could in a conference phone call involving Canuto in New York, Donahue in Tucson, Harbottle in Brookhaven, and me in Rochester.

The call was placed. Canuto told us he got to Rome on 15 June. Chagas had been there a couple of weeks. An official letter awaited him. It was signed by the Cardinal Secretary of State, second in command at the Vatican, and addressed to Cardinal Ballestrero. It gave him the final 'marching orders' concerning the procedures to be followed in dating the shroud.

Included was a letter from the secretary of state to Chagas thanking him and the academy for having organized the Turin workshop. The inference was clear—the Pontifical Academy of Sciences had done its job and now everything was in Ballestrero/Gonella's hands.

Canuto stated that the letter to Ballestrero from the Vatican was dated 21 May and it must have reached Turin in 5 or 6 days. Ballestrero had now had it for at least a month. He said "They have not communicated with us (meaning the Pontifical Academy) and we have not communicated with them. We don't know whether they have indicated to anybody what their decisions are." He said, however, that while Canuto was with him in Rome, Chagas called Tite over some point on carbon dating that needed clarification. Tite volunteered that he had been contacted by Turin and he knew about this decision of the Vatican. Chagas had replied to his letter from the secretary of state acknowledging the pope's order that his work was ended—an order he has to obey.

I said "Chagas has told me several times that the pope gave him the authority to carry out the carbon dating of the shroud". Canuto said he and Chagas had read the original letter of instructions from the pope to Chagas very carefully—the letter said very explicitly that Chagas was requested to organize the meeting in Turin. Chagas understood that in a wider context— if the academy organized the meeting, they would be in it for the whole process, but that is not the way the Vatican authorities finally chose to interpret it.

Harbottle then raised the point that a letter from the secretary of state to the cardinal in Turin must have been preceded by some communication between Turin and the secretary of state. Canuto said, "It's good to talk to more people because I did not want to volunteer that information, but since you are asking I have to give it to you. The letter from the secretary of state to Ballestrero begins by saying 'Dear Cardinal: We have received a letter from Carlos Chagas on 8 October 1986' [this is a letter that Chagas and I (Canuto) wrote to the pope upon returning to Turin from Rome in which we sent the protocol and everything] 'taking that letter into consideration *plus the letter of January 25 of Your Eminence*'. That means a letter that Ballestrero wrote to somebody at the highest authority of the Church." (So now we know for sure that Ballestrero requested the pope to send him a letter authorizing the revisions of the protocol and the pope did so through his secretary of state. These revisions were undoubtedly suggested to Ballestrero by his trusted science advisor, Luigi Gonella, and were just the ones Gonella wanted.)

Canuto went on to remark that the Vatican did not send Chagas a copy of this letter of 25 January from Ballestrero to the pope. He excused this on the grounds that it was a private letter between two cardinals.

Canuto said he then had assisted Chagas in writing another letter to the pope via the Vatican secretary of state. It described the size of the sample required for all seven laboratories, 0.025% of the total shroud area. There would be little savings in material by reducing this by a factor of two or so—the amount required was trivial in any case. The letter quoted from the article in *Radiocarbon* on the laboratory intercomparison tests carried out by the British Museum. It recommended that *several* laboratories be involved in the carbon dating of the shroud because of the outlier result obtained by one laboratory in that test. Canuto said he had checked the *Oxford Dictionary* to find out what the word 'several' meant. It meant more than two or three, so in the letter they stressed that more than two or three laboratories be involved.

It was now obvious, although never explicitly stated by Canuto, that 'the marching orders' included a reduction in the number of laboratories to be involved—probably to the two or three that Gonella had always favoured. Canuto said that neither he nor Chagas had regarded this as a disaster—they did not go out and buy a bottle of champagne to celebrate—but at least it meant that the whole thing was up and running. Harbottle added that the pope had given his authority for the shroud to be carbon dated and that was a big step forward. Canuto said, in the end, he and Chagas just congratulated each other. They did not understand why not seven laboratories—this was a worry because of the foreseeable reaction of the laboratories that were to be excluded. I said that a key question was who would make the decision as to which laboratories would be chosen to do the job. Would the choice be made by Professor Gonella? Canuto observed that the letter was addressed to Ballestrero and therefore Ballestrero, i.e. Gonella, would do the choosing.

Canuto went on to say that the laboratories could all, as a corporation, stick to the Turin Protocol and refuse to do it unless all its provisions were adhered to. We discussed how we might protest this decision without revealing that we were privy to the existence of the letter from the secretary of state to Ballestrero. We were worried that it would be widely assumed that Canuto had revealed the detailed contents to us even though he had not.

Harbottle said that he thought there was a way out. We could refer to the *La Stampa* article. It quoted Gonella as saying only two or three laboratories would be involved. We could always assume that was an authoritative statement and that article certainly was in the public domain. Harbottle said that we could use that as a basis for soliciting opinions on what to do next. We could privately tell the laboratories that there was more to it than that.

Canuto repeated that Tite had already received this information via some communication from the Vatican or more likely from Turin. Canuto said he had also spoken with Hall because there was a misunderstanding dating back several months that Chagas wanted to clear up. Hall was under the impression that Chagas wanted only two or three laboratories to be involved. Canuto said he didn't know how Hall ever came to that conclusion. Hall said that he thought that, in Turin, Chagas had indicated it to be so, and Canuto said "No. Chagas never wanted three labs, it was the other way around. Hall was happy to hear that. I did not tell him about this letter. He asked me if anything had happened in Rome—from which I deduced that he had not been contacted."

Harbottle remarked that we had enough to go on without explicitly referring to the letter. We had the information from the *La Stampa* article. He thought we could let the other laboratories know that it really had been decided that fewer than seven laboratories were to be involved. Canuto noted that the *La Stampa* article had been published on 27 April, almost a month before this letter was written. At the time of the Turin workshop, Gonella made it clear that he favoured using only two or three laboratories. Now he had the orders essentially from the pope that he had all along wanted and maybe we should wait until he made his move.

I argued that we should have a position before Gonella sprung this on us. I said that I would like to think that when Gonella told us which laboratories he had selected we would all stick together and say we wouldn't accept that. That might bring Turin to its senses. There are no other laboratories that, under these circumstances, would be willing to take on the job. I thought that it wouldn't hurt to call the various people in the other four laboratories to find out what their reaction would be to such an announcement from Gonella. We agreed that the letter from Cardinal Casaroli, the Papal Secretary of State, to Cardinal Ballestrero should not be specifically mentioned.

Canuto said "I got back a week ago and I knew that I had promised to call you. I felt bad that I couldn't call you because I gave my word to Chagas that we wouldn't volunteer anything, but neither would we lie. So it is only because Harry called me this morning, and I knew that Harry would call me, that I can now provide this information." Harbottle said "It was only a matter of time because I was going to call you also". (Maybe he was—but I pre-empted him.)

Since it now seemed clear that we were going to contact the other laboratories to see what their reaction was, I said, "Let me try this out on both of you [i.e. on Harbottle and Donahue]. Suppose the decision is for there to be only three laboratories and that you, Garm or Doug, get a letter from Gonella saying you have been tapped to be one of the three. What would your reaction be?" Harbottle said "Of course I would say no, I want to stick by the protocol". I went on "And Doug, what would you say?" Donahue replied "I wouldn't do anything without a lot more talking, I would want to stick by the protocol". I said I thought that is how the other people would feel also. It was certainly how I would have responded.

We finally ended this lengthy conference call by agreeing that I should contact all seven laboratories and tell them that the decision announced in *La Stampa*, that there would only be two or three laboratories involved, was going to be approved by the pope and the cardinal. We had information from a source I could not disclose, that this was the way it was going to be. What was their reaction to it?

I prepared a few notes to refresh my memory for the phone calls I was going to make to people at various laboratories. The notes read: I had just received information from a completely credible source that, as a result of a request by the cardinal of Turin to the pope, it is agreed that: (1) Pontifical Academy involvement in carbon dating the shroud ended with the Turin workshop; (2) Turin, i.e. Gonella, was in charge of the project; (3) fewer than seven laboratories would be involved—maybe as few as two or three.

The first person I contacted on 30 June 1987 was Woelfli in Zurich. I outlined the above information. I said one possibility was to compose another of these damned joint letters, in this case, directed to Cardinal Ballestrero. It would suggest that if there were going to be such major revisions in the protocol to which we had all agreed, including the people in Turin, that either they had to have another meeting with the people directly involved in carbon dating or we would just say to hell with it. I outlined a letter that the seven laboratories and the British Museum might send directly to the cardinal in Turin. Woelfli said he would agree to such a letter.

I next talked to Hall in Oxford. I asked him whether he had heard from Rome or Turin. He said that he had only heard from Canuto (as Canuto had told us). I gave him the same information I had given to Woelfli. He asked whether the British Museum would be present and I said it was not clear whether they would or not. I suggested that another joint letter be sent, this time addressed to Ballestrero. He immediately interrupted me to say: "Now look Harry, I suggest that you don't do anything. If you do anything to your enemies in Turin it will be curtains." I was taken aback by his use of the word 'enemies'. I had realized by this time that I could not list Gonella as one of my warm friends but surely not as an enemy! He said I must understand his point—I was not in the good graces of the people in Turin. I said I was proposing a joint letter, obviously not a letter from me alone. I had already talked to Woelfli in Zurich, and also to Donahue and Harbottle, and they all think that we cannot accept a change like this without another meeting. I then read him the suggested message to the cardinal, that ended with the phrase "the undersigned state that without such a meeting, a unilateral major modification of the Turin Protocol would not be accepted and the laboratories would withdraw their offer to be involved in the project to establish the age of the Turin Shroud." I added that Woelfli, Donahue and Harbottle were in agreement with this wording. He said that it sounded a bit threatening to him. He said he thought we should not threaten them in any way. He went on to say: "I think you should say that we believe that the meeting should be held and leave it at that and not say 'otherwise we will withdraw'. As soon as you start threatening people, you will only get their backs up." I agreed to omit that statement. It was clear that Hall wanted to keep his options open for Oxford to date the shroud.

I phoned Donahue. He said what had happened was almost certainly the work of the STURP committee. However, once the Vatican had made a decision like this it would be difficult to get it reversed. I replied that I could not help thinking that it was STURP's enmity toward me that was causing the problem. He said he did not think I was being singled out. He admitted that STURP liked me less than the others but what they really didn't like was our meddling. I told him that I had discussed a possible communication to Ballestrero with Woelfli and with Hall. Hall expressed concern about the perceived threat we were making if we said we would withdraw from the enterprise unless we had another meeting to decide on the changes that were being dictated—particularly the reduction in the number of laboratories. Donahue felt it might be considered a threat. He suggested a slightly different wording: "If the protocol agreed to by representatives of the seven laboratories is to be altered in a significant way then we should meet again before we can proceed any further." I agreed—it was essentially what Hall had suggested. I noted that I had the very strong impression from Canuto that he and Chagas expected us to write a letter along these lines and they would be amazed if we did nothing.

I contacted Harbottle and read him my preliminary draft of a letter to the archbishop of Turin. I told Harbottle that I had in mind that only representatives of the seven laboratories and the British Museum would be asked to sign the letter. Harbottle was worried about the part at the beginning where I referred to *La Stampa* and suggested there were other indications of changes in the protocol. He thought that was risky because it might compromise Chagas and Canuto. I told him that in discussing the matter with all the people I had called, I had been very careful to say that there were strong additional reasons to believe that the *La Stampa* article was correct. I could not reveal the source of this information but it was credible. I agreed with him that it was probably unnecessary to indicate to the cardinal that we had any information beyond what was contained in the *La Stampa* article. Harbottle also thought that we should make some mention of the fact that the cardinal, in his opening address at the Turin workshop, said that it was important that this whole procedure of dating the shroud be carried out in the proper way. It was to satisfy this desire on his part for a convincing and valid test that we drew up the protocol. This change was not in the best interest of doing the job correctly.

We decided that it would be useful to have Canuto translate the text of whatever we agreed to into Italian in order that the cardinal could read it directly. If it were written in English then it would be Gonella who interpreted it to the cardinal in any way he wished. We agreed that I would get a draft written, discuss it with the other labs by phone, and send it to Canuto for translation. He said he thought we might still be able to head Gonella off.

I then phoned Michael Tite at the British Museum. He was a bit cagey as to what he knew. Clearly he knew a great deal more than he revealed in this phone call. He finally agreed that if I were to send him a copy of the proposed communication to the cardinal that indicated there should be another meeting of the laboratories and the British Museum before major changes in the protocol were made, he would agree to sign it. If it were finally established that it had to be three, the laboratories themselves should decide which three.

On 1 July 1987, Canuto phoned me and asked excitedly: "Did you hear the good news?" I replied, "I don't have any good news, what's yours". Canuto replied, "He's coming, he's coming to the States". Despite my deep involvement with the reputed burial cloth of Jesus, I was confident it was not Christ's second coming Canuto was signalling. I asked: "Who is?" and he replied "Cardinal Casaroli!" He was coming to receive an honourary degree at some university or other. Canuto hoped that he would be able to at least have a phone conversation with him.

I discussed sending our message to Cardinal Ballestrero. It should be in Italian and we were hoping that Canuto would provide the translation. I read him the letter. It had been substantially rewritten from the version I had read to Harbottle on the phone and it incorporated Harbottle's suggestions that we quote from the statements that had been made by Ballestrero himself at the Turin workshop. It went as follows:

"Your Eminence, In your opening statements to the delegates of the Turin Workshop on Radiocarbon Dating of the Turin Shroud, you charged us with designing 'a valid and acceptable project for at last carrying out the radiocarbon dating of the shroud cloth'. You reminded us that owing to the 'unique and singular character of the object, such measurements certainly could not easily be repeated'. You asked us to devise a concrete operative proposal. Your Eminence, it was to satisfy this desire on your part for a convincing and valid test proposal that we drew up the protocol. A most important article in the protocol concerns the minimum number of independent measurements required to fulfill your charge of achieving a credible result. In our judgement that number involved measurements by seven different laboratories. We were therefore alarmed to read in the 27 April 1987 issue of *La Stampa*, a statement attributed to Professor Luigi Gonella, your science advisor on matters concerning the Shroud of Turin, that only two or three laboratories will be involved in the measurements. If that is indeed the case, you are risking the possibility that what may be the first and only chance to date the shroud will fail. The material removed from this precious object may be wasted. We urge Your Eminence before making a final decision on this question to reconvene a meeting of representatives of the seven carbon dating laboratories and the British Museum with your science advisor, Professor Gonella, to more fully apprise him of the dangers of modifying the Turin Workshop Protocol in this fundamental way. The protocol was carefully crafted to meet your charge that the results of the measurements be credible to the general public and to knowledgeable scientists alike. As participants in the workshop who devoted considerable effort to achieve your goal, we would be irresponsible if we were not to advise you that this fundamental modification of the proposed procedures may lead to failure."

It was our idea to send Canuto a copy of the cable and have him translate it into his marvellous northern Italian. This would bypass the need for Gonella to translate it with all the attendant potential for mischief. Canuto said as soon as he got it he would get to work on the translation.

On 1 July 1987 I express-mailed the letter to Canuto. I included the cardinal's address to the delegates in Italian as copied from the Turin workshop transcript. I thought that this would enable Vittorio to make sure that the phrases we were using in this letter did correspond to what Ballestrero had said at the opening of the meeting. I also sent a memo to the representatives of the seven carbon dating laboratories and the British Museum in which I enclosed a copy of the English version of my letter to the cardinal. I said that I hoped we could all agree to sign. The important thing was to get a fast response from everyone so that the translation by Canuto could be sent to the cardinal before he committed himself to this path of folly. I added that it might already be too late.

I called Canuto next day. He said that he had received the letter to Ballestrero just a few minutes ago. We decided that a copy should be sent to both the secretary of state at the Vatican, Casaroli, and to the Pontifical Academy of Sciences.

Canuto said he was impressed by the fact that, since it would be written in Italian, the cable would go straight to Ballestrero and he would read it. He might then call Gonella in for some interpretation, but the message would have sunk into his spirit. Canuto said that he would translate the letter, type it the next day and get it back to me by express mail.

That same day, Donahue and Damon called. They said that they would sign the cable to Ballestrero. Damon was appalled at the change in the protocol. He said that if Arizona were one of the laboratories selected, they would probably refuse. He said he saw no reason to change the protocol since it was a remarkably complete and satisfactory document.

On 3 July I managed to get through to Otlet. He was in favour of sticking by the protocol, especially since his management had agreed to it and he would have to go back to them if there were any substantive changes. He said I could list his name on the letter.

I tried calling Hall but was told he was on his way back from London and I should call back in a couple of hours. I phoned Tite directly after that and he agreed to sign. I phoned Woelfli and later received an electronic mail message from him confirming his agreement.

Later on 3 July 1987, I managed to contact Hall. He said that he had read the letter and was not very happy about it. He thought we ought to write a letter, but all this stuff about seven labs being a magic number was really open to criticism, why not 22 or some other number? I mildly remarked that it just happened to be the number of laboratories involved in the workshop and agreed to in the Turin Protocol. He said maybe it was okay to use the number seven. I said Tite, Woelfli, Otlet, Damon, Donahue and Harbottle had agreed to their names on the letter. He asked if I had talked to Tite recently? I said I just phoned him about an hour or two ago and he had agreed to sign. Hall then made a strange remark. He said, *"It didn't change anything then, hmmm"*. This led me to believe that Hall, like Tite, knew a little bit more than he was willing to disclose. He probably knew that Tite had some information from

Gonella—possibly about which labs were chosen and that one of them was Oxford.

I went on to say that the letter was being translated into Italian and that I would get a copy of this translation in the mail. I had in mind sending a cable to the archbishop with all our names on it and perhaps I would have it sent by Canuto—I was trying to respond to Hall's comment that he didn't think it should come from me. He said "Yes, I appreciate that Harry. I don't want to be a nuisance." I assured him he was not being a nuisance, the real nuisance was Gonella.

I said all of us felt there should be a meeting again of the principal participants—these being the seven labs and the British Museum—with Gonella. He said he entirely agreed with reconvening the meeting, he had no arguments against that at all.

He asked me to explain why I thought it might fail if there were only three labs instead of seven. Although I was bemused that he needed such an explanation, I said he needed only to look at the interlaboratory comparison paper the British Museum presented at Turin. If one took an average of three of those measurements, the outlier and any two others, one would get an answer that was several hundred years wrong. That was the kind of thing that could happen. I said it was not likely, but it could happen and using seven labs would increase the probability of identifying an outlier. I did not mean to imply that there had to be seven labs but seven of us were involved in the Turin workshop. The final protocol named seven. I just did not feel that any of us could arbitrarily say that it can be limited to any other number. I said that that was something that might be settled at another meeting, but I thought it was quite arbitrary that Gonella should decide that there would be only three. I wondered what his motives were? Certainly he was not motivated by saving cloth. The amount of cloth that would be used by seven labs would be trivial and it would not be much more than would be required by three. Hall said that he suspected Gonella did have other motives. I remarked that Otlet had said that he would have to go back to his management if there were major changes in the protocol, but that he was happy to sign because all it was saying was that we wanted to do what his management had already agreed to do.

Hall asked me if it were really true that Chagas was out of the picture. I said yes, Cardinal Casaroli wrote to Chagas making that clear and I thought it was a real pity. Hall said: "I agree, it is a real pity". Finally Hall agreed to have his name appended to the letter.

Around 5:30 on the 3rd of July Canuto phoned. He had sent his Italian translation of the letter to my home address by US Post 'next day' Express. He said Cardinal Casaroli would be arriving in New York on Monday 6 July and would be staying at the Holy See Mission to the UN. Canuto would be attending a dinner in his honour on Wednesday evening. He suggested that I phone Monsignor Franco at the Holy See Mission and say that I wanted to send Casaroli a copy of the cable to Ballestrero. Franco will probably tell me to send it to the Holy See Mission. In that way Canuto could have a chance to discuss the new shroud developments with Casaroli. He said that a copy should also be sent to Chagas at the academy headquarters in the Vatican.

The next day, 4 July, about 10 am, I phoned Canuto and said that I had called the US Post Office and got no answer. I doubted whether Express Mail would be delivered today—after all, 4 July is a holiday in the USA. He said that his son had been told by the Express Mail Office in New York City that it would be delivered 4 July, so I was to just stand by. I then called Monsignor Franco at the Holy See Mission and he said it would be okay to telex the letter to Cardinal Casaroli care of the Holy See Mission and he gave me the number. About 2:45 pm the Rochester main post office called to say that some mail had arrived for me so I drove there to pick it up.

Before sending Canuto's Italian translation off by cable, I felt I really had to try to get permission from Duplessy to let me add his name to the cable. On 6 July 1987 I finally managed to contact him. He said that he would be happy to have his name on the cable to Ballestrero. He said that we had agreed on seven laboratories at the Turin workshop, and he still stood by that. I gave him the information that Cardinal Casaroli was in New York and that Canuto hoped to be able to discuss the matter with him.

The cables of the Italian version of the letter to Ballestrero were sent on 6 July. Shirley and I went downtown to the Cadillac Hotel, the local office of Western Union—a rather seedy hotel in a rather seedy part of town. The operator there typed the document in Italian into the system. It was sent to Donahue, Duplessy, Hall, Harbottle, Otlet, Tite, and Woelfli. Copies were sent to Casaroli, both to the Vatican and the Holy See Mission, and to Chagas at the Pontifical Academy of Sciences. The principal copy was sent to Cardinal Ballestrero.

On 9 July 1987, Canuto called. He had had lunch with Cardinal Casaroli and drove with him to the airport. Regrettably, no opportunity had arisen to discuss shroud matters with him. I was disappointed and wondered to myself how Canuto, with all the time he spent with Casaroli, had failed to discuss the subject. Perhaps he was too overawed by His Eminence's eminence. After all, in the Vatican hierarchy, Casaroli was number two after the pope.

On 19 August 1987, I received a letter from Hall, dated 12 August, that was most disquieting. He recalled our telephone conversation concerning what he described as my 'broadside' to Turin (the letter in Italian to Ballestrero) he had told me he considered unwise. He said he had agreed to sign it as a friendly gesture. He had since learned it had displeased the archbishop who was now probably laying various mystical punishments on our heads which he has the power do. He stated that from now on he and Hedges intended to distance themselves from the two camps, me on the one hand and Professor Gonella on the other. He thought that any further hectoring would only prolong the decision. He hoped for a positive decision for the shroud dating sometime in the future. Meanwhile he intended to keep quiet and await developments from Italy. He hoped this would not cause offence but he wanted to make his position clear.

Hall was displaying a most un-British spirit—he was breaking ranks—he was letting the side down! My guess was that Hall sensed, if he did not already know, that if he did not further ruffle any feathers, Oxford would be one of the laboratories Gonella would select.

On 21 August 1987 I replied to the above letter as follows:

"Dear Teddy: Thank you for your candid letter of 12 August 1987. It has always pleased me that the modest role I played at Rochester along with my collaborators Ted Litherland and Ken Purser over ten years ago was the genesis of the AMS facility in your laboratory at Oxford, our friendship, and the beginning of the shroud adventure. One cannot take offence at frank words from a friend.

"Please permit me to cavil mildly with your description of our joint cable, which you co-signed apparently only in a spirit of amity, as a 'broadside'. I should have thought a more accurate expression in naval parlance would be 'firing a shot across the bow'. It was couched in the most courtly, civil, and almost obsequious terms. Although it may have displeased Professor Gonella I have no indication through my Vatican channels that it displeased the cardinal. If it did one is hard put to understand why.

"Let me hasten to assure you that my 'hectoring' as you call it is directed toward STURP and only peripherally toward Professor Gonella to the extent he champions STURP's cause. It is certainly not directed to him in his capacity as the cardinal's science advisor. By all accounts, however, he is not held in particularly high regard in that capacity outside Turin. He is unfortunately still *the* power in Turin as far as the shroud is concerned. I have had almost ten years' experience with STURP and regard them as a pack of religious zealots who could really queer the pitch for carbon dating unless they are held at bay. I fear the cold and malevolent eye of Mr Lukasik much more than your suggested mystical imprecations of the cardinal. I have received recent information that STURP's influence is on the wane and high bloody time I would say.

"I hope I can be as frank with you as you were with me in what I am about to write. In what follows it will be useful to recall three letters which appeared recently in *Nature*. I attach copies for your convenience. Before sending my letter in answer to Denis Dutton I obtained the approval of Professor Chagas who read it and who, at the time, was still in charge of the shroud dating project. You will note that, in my letter, I stated that the protocol agreed to by all parties at the Turin workshop will be published in the November issue of *Nuclear Instruments and Methods in Physics Research*, section B, and so it will. This paper authored by me was refereed by Paul Damon. The protocol which formed the bulk of the paper clearly provides all the answers to Dr Dutton's concerns.

"At Trondheim in June 1985, at the informal meeting I organized which culminated in the Turin workshop, the question was raised concerning the participation in the shroud dating project by both the British Museum and Oxford. This was because of the dual role you play as a member of the Museum's Board of Directors and as head of the Oxford AMS laboratory. I and a majority of the people present discounted this as a problem. However, in view of Dutton's letter to *Nature*, and I suspect this paranoia is rather widely shared, if the number of carbon dating laboratories were reduced from seven to three, I would be seriously concerned as to how the validity of the dating would be perceived by the general public if both Oxford and the British

Museum were involved. I think we all agree that the British Museum must be involved and apparently now so does Turin. If that eliminated Oxford it would be a great pity. If, on the other hand, all seven carbon dating laboratories were involved it would be hard to argue that any hanky panky had taken place between any laboratory and the British Museum.

"I personally regard the participation by both the British Museum and Oxford as important. If the number of laboratories is unnecessarily and inexplicably circumscribed, arguments might be raised for keeping the British Museum in and dropping Oxford. It would then be pretty difficult to decide which three dating laboratories should be selected

"Why the hell should Turin be permitted to put us in this 'Sophie's Choice' situation? Do any of us so lust to have our laboratory involved that we are willing to behave in unseemly ways? I suggest we continue to stick together. There is no earthly reason why three laboratories are better than seven. They cannot date it without us and who of us really cares if it is ever dated? I suppose I should care the most since I have put so much time and effort into the enterprise. What I do care about is that if it is done at all it be done right. The agreement on procedures we reached in Turin is likely to be as close to the right way as we are ever likely to invent.

"I do not agree with you that the original 'protocol' is dead but I do agree that we should now all 'keep quiet and await developments from Italy'. I just do not believe you are willing to break ranks at this critical time as your letter seems to suggest. With warm personal regards."

After receiving Hall's letter, I was still very annoyed because it really did mean he was breaking ranks and, I suspected, for self-serving reasons. I decided to send a letter to all of the lab people and did so on 24 August 1987. It was sent to Duplessy, Tite, Donahue, Damon, Hall, Harbottle, Otlet, Woelfli and Canuto. In it I outlined the contents of Hall's letter, quoted some parts of my letter in reply and added some conclusions I had reached, as follows:

"I had a letter the other day from Teddy Hall who reiterated a view he had expressed at the time that it was unwise to send our cable to Turin although he agreed to sign it as a friendly gesture (to me I assume). He said he had indications that it displeased the cardinal. He chided me for hectoring Professor Gonella to the point of creating two camps—Gonella's and mine. He said that he and Robert Hedges intended to distance themselves from this concept. I should note here that there has been absolutely no interaction between me and Professor Gonella since the end of the Turin workshop except for a phone call I made to him on 10 April 1987 to find out what progress was being made on carrying out the Turin workshop agreement." I reiterated what I had said in my letter to Hall about 'hectoring' STURP and not Gonella.

"It got me to thinking, however, about the decision which appears to be final that only three of the seven laboratories will be involved in radiocarbon dating the shroud cloth. . . .Although we know from the *La Stampa* article the decision was taken some time in April we have not yet been officially notified. This indicates that our 5 July 1987 cable may at least have caused some delay. . . .However, it is unlikely to have caused any change in the decision to reduce the seven laboratories to three. Whether we will be invited to accomplish this ourselves (shades of Sophie's Choice) or whether it will be done by fiat by Turin has not yet been revealed to us. . . .It seems to me whether we are asked to select three labs or whether Turin imposes the choice on us, we should stand by the agreement reached in Turin.

"I believe that is the best way to avoid the suspicions raised by Denis Dutton in his letter to *Nature* (**327** 10, 1987). I am sure there will be similar feelings of paranoia on the part of many others if only three labs are involved.

"For example, since the British Museum will certainly play a role, will this compromise Oxford because Teddy Hall has close connections with both? If the Abegg-Stiftung is involved does that raise questions about the participation by Willy Woelfli's lab? If Rochester is eliminated because I am allegedly 'hectoring' Professor Gonella does that mean Gif-sur-Yvette and Arizona must be the two AMS labs (pretty good choices I would say) to maximize the international distribution and thus decrease the chance of collusion (this would then favour Harwell over Brookhaven)? And so on.

"The only way to avoid this idiocy is for all of us to stand by the Turin Workshop Protocol. All seven labs should participate, along with the British Museum and the Abegg-Stiftung. I also feel very strongly that we should continue to benefit from the participation by Professor Chagas as President of the Pontifical Academy of Sciences; the role described in the protocol. Representatives of the seven labs must also be present during the sample taking in

Turin and personally receive the samples and controls for delivery to their labs. If Turin intends to modify any of these protocol provisions I believe none of us should accede. I suggest in such circumstances Turin be invited to find three other carbon dating laboratories which would be willing to take on the task. Such a stance should not be taken as a threat to Turin but rather as an act of prudence and responsibility.

"I would be pleased to know whether you agree with me or not. This letter is being addressed to the heads of the other six laboratories and the British Museum. Yours sincerely."

I began to get replies to my August letters. One was from Teddy Hall, dated 3 September. He thanked me for my 21 August letter. He said his main contention was that the cardinal would make the decisions and nobody else. We have made it plain to him what it is we want but he is not obliged to go along with us. Any one of us are free to refuse if we do not think the conditions are correct. Hall said that he, for example, would think hard if less than four labs were involved. (He clearly did not have to think too hard when it was decreed that only three labs would be involved—one being Oxford. He accepted with alacrity.) He noted that the cardinal can presumably commission anybody to do a measurement, as hundreds of other samples have been commissioned, some very important ones. Some of us don't even think the shroud is very important, except for the public interest in the result. Hall thought that we all agreed if STURP ran the show we should all pull out. He did not wish to break ranks in any selfish way and indeed was in no position to do any such thing since he had no idea what the archbishop and his advisors had in mind. He thought it would be ironic if Oxford were disbarred from participating due to his British Museum connection. He claimed that the British Museum was his idea, and stated that Mike Tite needed quite a lot of persuading and his fellow trustees even more.

I did not know what he meant by this last statement. It was not his idea to have the British Museum participate in the actual shroud dating. Hall had said he was not even present when his fellow trustees of the British Museum agreed to participate, so he could have played no role in persuading them. Furthermore it was Otlet's idea, not Hall's, to have the museum play a role in the interlab comparisons. Hall merely conveyed the request to Tite.

Woelfli, on 3 September 1987, replied to my letter of 24 August.

He said that a couple of days earlier he had received a phone call from Gonella, informing him that our cable to the cardinal had not been so well received and even might endanger the carbon-14 project. Gonella also had made it very clear that in any case Turin would decide how many and which laboratories would get samples. Woelfli said that Gonella had sounded quite annoyed when Woelfli told him that he still favoured the procedures described in the Turin Protocol and that any modification of it would require a new meeting of the whole carbon-14 community involved. He went on to say he did not know what would happen now. Gonella had only told him that the final decision on procedures would certainly be delayed now. Woelfli wrote that he didn't care at all what happened but that he was most interested in the reactions of the other labs. He said he looked forward to hearing from me soon.

On 3 September I received a letter from Otlet. He wrote that he was sorry to hear of my exchange with Teddy Hall. He had heard from a colleague that Hall had caused great merriment in a recent lecture he gave in the next town, Abingdon, only seven miles away from Harwell. Hall had ridiculed the Harwell small-counter method, the time it took to make a measurement and the size of the sample required, which he described as a whole pocket handkerchief piece of cloth. (Taking such a large sample had been denied more than forty years ago to the inventor of carbon dating, Willard Libby. Hall knew full well that Harwell needed a sample not much larger than Oxford required—a thousand times smaller than Libby would have needed.) Otlet went on to say that he had to admit he was a little uneasy about my approach to the cardinal. However, he reiterated his position that any substantial change in the protocol would not be acceptable to him and he had written to Chagas saying so directly. He said he would be interested to know whether I had had similar assurance from anybody else.

Sox called me sometime during September 1987. He quoted Ballestrero as saying "the shroud is my crown of thorns". He had learned from Rinaldi that if the carbon date came out right, the authorities in Turin had a plan to display the shroud in an edifice resembling Napoleon's tomb. He claimed they had actually sent an architect to Paris to study the tomb.

I received a letter from Tite dated 11 September 1987, thanking me for my letter of 24 August. He said "I certainly agree with many of the points you make in that letter, however I do not feel that it would be particularly useful to discuss or comment on the matter further until we have received some specific proposals from Turin. Like you I certainly hope we hear something soon."

There was a very brief letter dated 15 September from Paul Damon, with a copy to Doug Donahue. He said, "In reply to your letter of 24 August it is my opinion that we should stick to the original protocol or not participate. Doug agrees. Cordially, Paul."

I now had statements from the heads of all seven laboratories that they would not break ranks, although Hall had been the most equivocal. The three that were eventually chosen, however, finally agreed to accept Gonella's dictum that they renounce the Turin Workshop Protocol. Hall (Oxford) had said he would think very hard if the number of laboratories to be involved were to be less than four. Woelfli (Zurich) had said that any modification of the Turin Workshop Protocol would require a new meeting of the whole carbon dating community involved. Paul Damon (Arizona), writing for himself and Donahue had said that we should stick to the original protocol or not participate. I surely could be excused for assuming that the ranks still held firm. Alas, such was not to be the case.

Chapter 8

Dating Labs Announced, Unity Attempts

On 19 October 1987, Otlet phoned from Harwell and said he had received a letter from Ballestrero announcing that Arizona, Oxford and Zurich had been selected to date the shroud. The letter had been sent to all the participants. My first reaction was to say that the exclusion of Rochester meant I should probably take a course in diplomacy. I admitted that I was most disappointed (a monumental understatement) and he said that he understood. I was genuinely concerned that, after what had happened in the British Museum's interlaboratory comparison, three was too few a number to statistically identify an outlier result. Aside from my deep disappointment that my laboratory would not be involved in dating the shroud, I wanted the job to be done right. Using only three labs was not the right way to do it.

Otlet sent me the archbishop's letter by fax. It was dated 10 October 1987, and went to all the participants in the Turin workshop.

"Dear Sirs: [Shirley got the same letter with the same salutation as, I assume, did Madame Flury-Lemberg]. At the end of May I received positive instructions from the Holy See personally signed by the Cardinal Secretary of State on how to proceed to the radiocarbon dating of the Shroud of Turin.

"These instructions agree to the main line of the proposal put forward at the Turin workshop of last year, but do not accept a few items. In particular they direct that no more than three samples be taken to be used for measurements by different laboratories. As for the measurements, the instructions agree to the suggested procedure, i.e. to use the method of blind testing

213

with control samples, to apply to the competence of the British Museum for supplying such samples, and for cooperating with me in the certification of the Shroud samples, and to entrust to the competence of the same British Museum and of the Institute of Metrology 'G Colonnetti' the statistical analysis of the measurements results.

"As a consequence, in the first place I wish to express my thanks to all who participated in the Turin workshop with generous availability, even though I find myself unable to take advantage of the competence of all participants, as it was in the wishes of the meeting.

"The choice of the three laboratories among the seven which offered their services was made, after long deliberation and careful consultation, on a criterion of internationality and consideration for the specific experience in the field of archaeological radiocarbon dating, taking also into account the required sample size. On this criterion the following laboratories are selected:

Radiocarbon Laboratory, University of Arizona
Research Laboratory for Archaeology, Oxford University
Radiocarbon Laboratory, ETH, Zurich.

"The operations for taking the samples have to be presided (over) by myself, in my capacity as the Pontifical Custodian of the Shroud. H. E. Professor Carlos Chagas, President of the Pontifical Academy of Sciences, will be invited to be present at the operation as well as at the eventual final meeting, as my personal guest, in consideration of the collaboration he gave in working out the project. The instructions from the Holy See do not deem it necessary for representatives of the measurement laboratories to attend the sample-taking operation.

"The decisions took more time to be worked out than originally wished, owing to the situation without precedents created by a number of competing offers tied into a rather rigid proposal, and also by (the) initiative of some participants in the workshop who stepped out of the radiocarbon field to oppose research proposals in other fields, with implications on the freedom of research of other scientists and on our own research programmes for the Shroud conservation that asked for thorough deliberations.

"Besides, when the competent Authorities advised me they deemed we ought to proceed with three samples, a concerted initiative was taken to counter the decision, with the outcome of a telegram sent to H. E. the Cardinal Secretary of State and myself by some participants in the workshop, a telegram where the meaning of my introductory words at the workshop was heavily misinterpreted.

"After further deliberation and scrutiny of the situation with the Cardinal Secretary of State, we are now proceeding on the already decided terms, that I was just going to write you when I received the above quoted telegram.

"In consideration of the great attention from the public and the press that all of us know that this measurement is attracting, it seems to me worthwhile to stress again what I said in my opening address at the Turin workshop, about the purely scientific character of this enterprise, which does not mean to, nor could, address any issue related to the death and resurrection of Jesus Christ. Nor do I mean with this analysis to charge the laboratories that have been selected with the task of 'authenticating' the Shroud of Turin: the analysis is strictly meant to ascertain the radiocarbon date of its cloth, as an objective datum of the highest importance for evaluating the complex issues of its authenticity and conservation.

"As we move on to the executive phase of the project of radiocarbon dating the Shroud of Turin, I would like to thank again all those who brought positive contributions to it, and to offer my heartfelt good wish to those who will undertake it, trusting that they will carry it out with utmost scientific rigour in order to add this important objective datum to the scientific quest that has long been growing on the illustrious image entrusted to my stewardship. Anastasio Cardinal Ballestrero, Archbishop of Turin, Pontifical Custodian of the Shroud of Turin."

It was altogether a most remarkable document and I had a number of immediate reactions to it. The first was that the President of the Pontifical Academy of Sciences was invited to be present as the guest of the Cardinal, which meant that he would play no role at all. The second was the statement that the laboratory representatives need not be present at the sample taking—that was absolutely outrageous. It was also clear that the cable we had sent to Ballestrero had really annoyed him intensely. In particular, he stated in this letter that the meaning of his introductory words to the Turin workshop were 'heavily misinterpreted'. One had to concede that only he could know what he meant to say in his introductory remarks but I think we correctly reported what he actually did say. It was interesting that he had been just about to send his letter off when he got our cable and that caused a delay.

His thinly veiled accusation that we were attempting to prevent STURP from carrying out its scientific investigations was quite accurate. It was not a question of interfering with anyone's scientific freedom. It was just that several of us felt very strongly that the dating of the shroud must not involve STURP and had to precede any sort of measurements that were done by STURP. I certainly did. There was also the concern that some of STURP's tests might damage the shroud. At no time had I ever stated that STURP should be prevented from making additional measurements after the carbon dating was effected. I had never before been directly or indirectly chastised by a cardinal and I resented the falseness of his charges.

On 19 October 1987, I phoned Donahue. He said that he had been really taken aback by this letter and he thought that they would probably not go along with it. He suggested that we should meet in Yugoslavia at the *13th International Radiocarbon Conference* to be held in Dubrovnik, 20–25 June 1988. I said that I would try to contact Woelfli.

Harbottle called me the next day and said that he was going to contact National Public Radio, *The Skeptical Inquirer*, Dutton, and whoever else he could think of to rail against this decision. He said that he would call Canuto and perhaps Chagas. He thought that Chagas could get through to the pope. He regarded the whole thing as a real disaster.

The next morning, I called Harbottle again. He said he had talked to Canuto and Canuto agreed that Chagas should have one more shot at the pope to persuade him to tell Ballestrero that he is getting bad scientific advice. I suggested that if we must go public, we should do it in some joint fashion involving Damon and Donahue as well as Harbottle and me. We could, for example, issue a joint press release. If it just came from Harbottle, from me or from both of us, it would sound like sour grapes.

Canuto called shortly after. He said that he had talked to Donahue and that Donahue was in pain—he thought that a new meeting should be held. Canuto suggested that we send an open letter to the pope via Chagas, along the lines that the workshop in Turin had involved seven labs, they had produced a protocol providing the best advice that could be obtained on how to date the shroud, and that Cardinal Ballestrero was getting bad advice from his scientific advisor, Professor Gonella. I proceeded to prepare such a letter.

On 26 October I phoned Woelfli in Zurich. He had just returned and read Ballestrero's letter but had not had time to digest it. What bothered him even more than the reduction in the number of labs, amazingly enough, was the fact that the laboratory representatives would not be allowed to witness the sample removal. Perhaps he had an intense desire to see the shroud first hand. On the other hand he was probably, and quite rightly, concerned that if he and representatives from the other two labs were not present at the time the samples were removed from the shroud and could not receive them then and there, they might not know for sure they actually had shroud samples. He reacted quite strongly against a press release. He said he would like to receive a draft of the proposed press release and also the letter that was proposed to be sent to the pope and would decide then on whether he was in agreement.

That same day, I sent a telex message to Duplessy and asked him to phone me at the NSRL or at home. Duplessy phoned the next morning and said that some time ago he had received a phone call from Gonella. He had made it clear to Gonella (he had spent an hour on the phone with him) that he did not want any change in the protocol and he described Gonella as being very unhappy. Duplessy said that he would sign a letter to the pope if all the other labs did as well, but he did not want to be involved in any press contacts. I had been wondering, from time to time, exactly what Duplessy's position vis-à-vis the shroud was. As I got to know him better through these various contacts, it became clear that he thought it very unlikely that it was Christ's shroud.

On 29 October 1987, I prepared a final version of a memo for the seven radiocarbon dating labs. It was sent to them either by express mail or by electronic mail.

"Some of us have been persuaded, as a result of the 10 October letter from the Archdiocese of Turin, that a direct appeal to the pope, preferably through Professor Chagas, is now in order. Such a letter would only have a chance of being effective if it were signed by at least six of the seven laboratories and I would only send it in that circumstance. If, for whatever reason, this proposed appeal fails it is hard to see how this unilateral rejection of the Turin Workshop Protocol by the Cardinal of Turin can be concealed from the world press; hence the proposed press release. Please let me know as quickly as possible whether you would be willing to sign the attached letter to the pope."

Attached was a draft of a letter to Chagas that included the draft of my letter to the pope and a press release. I noted that we proposed to resort to the latter only if our appeal to the pope failed. I said that we hoped Chagas might be willing to endorse the letter to the pope and to deliver it to him in person. I listed the changes in the Turin Workshop Protocol being proposed by Ballestrero, clearly on the advice of Professor Gonella.

1. Five AMS and two small-counter laboratories reduced to three AMS laboratories.

2. No independent textile expert designated to remove the shroud samples.

3. Laboratory representatives not permitted to witness shroud sample removal.

4. No suggested involvement by laboratory representatives in the final data analysis.

5. No official involvement by the Pontifical Academy of Sciences at any stage. Professor Chagas invited to participate merely as a guest of the Cardinal of Turin.

The draft letter to the pope read as follows:

"Your Holiness: Following your specific instructions, representatives of scientific laboratories specializing in the technique of carbon dating small samples met in Turin on 29 September–1 October 1986, to discuss the protocol to follow should you permit the dating of the Holy Shroud of Turin. The workshop was held under the joint sponsorship of His Eminence Cardinal A Ballestrero and of the Pontifical Academy of Sciences. At the end of this workshop a detailed protocol was arrived at that was agreed upon by all participants. The two main guiding principles were:

1. Removal of a minimum amount of Shroud material. A total of $12\frac{1}{2}$ square centimetres (less than 0.03 per cent of the total surface of the Shroud) will suffice for all the laboratories.

2. Absolute scientific credibility. It was unanimously decided that, to improve the statistical credibility of the analysis, a minimum of seven laboratories should perform the dating. It is important to note that they use different techniques.

"In a letter dated 10 October 1987 to all the workshop participants, Cardinal Ballestrero has ordered substantial modifications to the original protocol. In particular, the number of laboratories is reduced, without any explanation, to three. This change was made in spite of a cable sent on 5 July 1987 to Cardinal Ballestrero with a copy to the Secretary of State, Cardinal A Casaroli, signed by all seven laboratories and the British Museum, in which the dangers of such a change were clearly spelled out.

"It is our collective impression that Cardinal Ballestrero has received very unwise scientific advice. The proposed modifications will confirm the suspicion of many people around the world that the Church either does not want the Shroud dated or it wants to have it done in an ambiguous way. The procedure that the Cardinal of Turin is suggesting is bound to produce a result that will be questioned in strictly scientific terms by many scientists around the world who will be very skeptical of the arbitrarily small statistical basis when it is well known that a better procedure was recommended. Since there is great world expectation for the date of the Shroud, the publicity resulting from a scientifically dubious result will do great harm to the Church.

"We respectfully urge Your Holiness to persuade the Cardinal of Turin that the scientific advice being given to him is not shared by the world experts in this field. He should be urged to seek the advice of the eminent scientific organization expressly created to advise you, namely the Pontifical Academy of Sciences that enjoys the respect of the scientific world at large.

"Rather than following an ill advised procedure that will not generate a reliable date but will rather give rise to world controversy, we suggest that it would be better not to date the Shroud at all".

The fourth enclosure was the proposed press release. It outlined the events up to Ballestrero's rejection of the Turin workshop agreement and his selection of only three labs to carbon date the shroud. It was an expanded version of the proposed letter to the pope. The concluding paragraph read:

"The new procedures suggested to the Cardinal of Turin and that he has now embraced, will, if implemented, yield a result for the date of the Shroud that will certainly be vigorously challenged by the world scientific community for their flimsy statistical basis. We urge the Cardinal of Turin to seek scientific advice from an unimpeachable source that was available to him from the very beginning, but that he chose to ignore, namely the Pontifical Academy of Sciences, which enjoys worldwide respect in the world scientific community. Only with the best advice of world experts on carbon-14 dating can a scientifically credible date for the Shroud of Turin be arrived at."

I awaited reactions from my colleagues.

On 2 November 1987, I received a call from Canuto. He told me that he had talked to Chagas over the weekend. Chagas would be in Rome until 23 November. Chagas said that when he read the 10 October letter from Cardinal Ballestrero, he felt as if he had been kicked in the stomach. He wondered what other peoples' reactions were. Canuto told him about the letter and the news release that we were proposing. Chagas thought that a letter to the pope was a good idea and he would probably endorse it and deliver it. Chagas particularly wondered what Woelfli's position was. Canuto said that he thought that Donahue was the key player. If he pulled out now, the US, where the new technique was invented, would be out of the action and that might have some impact on Turin.

I then called Woelfli. He said that he would be glad to sign the letter to the pope on the condition that both Oxford and Arizona sign. He said that he had now received a letter from Ballestrero specifically inviting him to participate (a similar one was sent to the other two labs). He said that his impression was that they should answer the invitation jointly. He thought that Oxford would certainly be prepared to go ahead. He said that he would very much like to hear from Donahue. He would like to coordinate the answer to the cardinal and he thought that it would be best for the labs to agree not to participate and then to sign the letter to the pope.

That same day Donahue told me that he favoured the three labs negotiating with Ballestrero and that if that were not successful they should then withdraw. He personally thought that the approach to the pope was fruitless and might mean that the shroud would never be dated.

Harbottle called and said he liked the material that I sent him. He agreed that we should just wait until the three labs decided what they were going to do.

On 3 November 1987, Hall told me that he would not sign the letter to the pope. He thought that Ballestrero would consider it blackmail. He said that he did not know how the other two labs would respond to the letter from Ballestrero, but he assured me that he would not go it alone. He said that if all three labs got the same result then of course everything would be fine, but he agreed there was some risk. He said that he personally thought that Gonella went from seven labs to three to get at me. He thought representatives from the labs must be at least in the next room when Tite supervised the cutting, and that they should receive the samples right there and then. He was particularly concerned that the British Museum be protected against the charge that Tite substituted samples. That charge could be made if there were no witnesses other than Turin authorities when the sample was taken under Tite's supervision. (In the event, that was what happened and such a charge was later made.)

I was not surprised that Hall would not sign the letter to the pope. He had come to believe that any protest we made to the way the Vatican/Turin were proceeding and the suggestion we might withdraw from participation would be regarded as blackmail. It might lead to the project being cancelled. There was no question he wanted his laboratory to be involved. The Ballestrero/Gonella decision meant Oxford would be involved. Why take any risks?

I then talked to Donahue. He said he was composing a letter to be sent jointly with Oxford and Zurich to Ballestrero saying that they were upset with the new procedures and would probably appeal to him to raise the number of labs to five or six. In any case, the letter would say they were unwilling to go along with three. He said if Arizona were to be included and Rochester were not, he would involve me as a consultant or in some other capacity.

On 4 November Canuto told me he had been on the phone for an hour with Peter Rinaldi and that Rinaldi was very unhappy about the cable we had sent to the cardinal. On the other hand he had not known about the article in *La Stampa*. Rinaldi said that the decision to use only three labs was made long before the Turin workshop. That, of course, was the way Gonella had been talking from the beginning of the workshop. That would scuttle Hall's notion, which I shared, that Gonella had decided on three because of the way I had antagonized him during the workshop.

I phoned Donahue that same day and told him that I thought he should tell Ballestrero that Arizona wanted to stick to the protocol. He said that he did not want to sign the letter to the pope and then he suggested that I call Damon. I did and Damon said that he also did not want to sign the letter to the pope. That was really not too surprising. If they were, indeed, preparing a reply it would be senseless for them, at the same time, to sign a letter to the pope. He said Doug's preliminary version of the letter would say that Arizona preferred that the original protocol be followed. If Ballestrero refused, then he did not know what they should do.

On 5 November 1987, I received a bitnet message from Arizona. It read: "Harry: This is a draft letter we propose might be sent from the three laboratories. Doug." He asked that I reply by bitnet. The letter read:

"Your Eminence: We have received your letter of 10 October 1987, and we are honoured to have been selected to participate in the determination of the age of the cloth of the Shroud of Turin. However, we are concerned to learn that a decision has been made to limit the number of participating laboratories to three. We are in agreement with the conclusions reached at the workshop held in Turin in September–October 1986, that is: 'a minimum amount of cloth will be removed which is sufficient to (a) insure a result that is scientifically rigorous and (b) to maximize the credibility of the enterprise to the public. For these reasons, a decision was made that seven laboratories will carry out the experiment...'

"We believe that reducing the number of laboratories to three will seriously reduce 'the credibility of the enterprise' which we are also anxious to achieve. As you are aware, there are many critics in the world who will scrutinize these measurements in great detail. The abandonment of the original protocol and the decision to proceed with only three laboratories will certainly enhance the skepticism of these critics.

"While we understand your desire to use a minimum amount of material from the Shroud, we believe that the increased confidence which would result in the inclusion of more than three laboratories in the programme would justify the additional expenditure of material. Although improvements in statistical errors resulting from including more measurements might not be great, the possibility of the occurrence of unrecognized non-statistical errors would be substantially reduced.

"For example, if only three laboratories participate, and one of them obtains a divergent non-understandable result, the entire project could be jeopardized, but if results from a larger number of laboratories are available, a divergent result could be more easily recognized as such and can be treated appropriately in a statistically accepted manner. Clearly it is the reduction of unrecognized non-statistical errors in measurements that leads to increased confidence in the final result.

"We would very much like to take part in the programme to determine the age of the cloth in the Shroud, but we are hesitant to proceed under the arrangement in which only three laboratories would participate in the measurements. We urge that the decision to change the protocol of the Turin workshop and to limit participation to only three laboratories be given further consideration. Respectfully..."

This letter spelled out in the most transparently unambiguous way the reasons for having the measurements made by more than three labs. It would add little or nothing to the statistical accuracy of the final result but it would provide a remedy for a rogue result by one laboratory as it had in the case of the British Museum's interlaboratory comparison. On 6 November I read Canuto the copy that I had received of Donahue's proposed message to the cardinal. Canuto's reaction to the letter was that it should be more specific as to what the three would like the cardinal to do, for example, hold another meeting. On the whole, however, he thought it was good. He said that he had had another long conversation with Rinaldi and that Rinaldi was coming around to our side. No one in Turin believed that the three labs would do anything but leap at the chance to do the job. However, it was gradually dawning on Rinaldi that this was not so and that the labs are genuinely concerned that the job must be done right and that this was not the way to do it.

In the issue of *Science News* of 7 November an article appeared under their Science and Society section titled 'Shroud Dating Isn't Ironed Out'. It reported that a stubborn wrinkle had developed in plans to date one of Christendom's best known relics, the Shroud of Turin. When the carbon dating experts met in Turin with Church officials last fall, they specified that six postage stamp sized samples be taken from the shroud to be tested in seven labs. The pope had apparently agreed to this protocol in February and had told the archbishop of Turin that the testing could commence. However, in an April interview in the Turin newspaper, the archbishop's science advisor indicated that not more than three samples could be taken. Only three laboratories, Arizona, Oxford and Zurich, would be involved, according to Harbottle who was listed as a spokesman for STURP. No reason for these changes was given.

On 17 November Donahue told me that the translation of their letter into Italian made by Canuto would be express mailed to Hall tomorrow, then to Woelfli and then to the cardinal. Canuto phoned me just after this call. Rinaldi reported to him that Gonella had talked to STURP's director, Thomas D'Muhala. He told D'Muhala that he would back off the decision for three labs if the three actually did stand firm in their opposition.

On 25 November an editorial appeared in Rochester's *Democrat and Chronicle* headlined '10 YEARS WAITING AND STILL NO

SHROUD'. It was the first time my name ever appeared in an editorial. It read:

"A decade or so ago, Harry E Gove was excited about his chances of discovering once and for all the real date of the Shroud of Turin. With only a tiny piece of the shroud, believed by some to be the burial cloth of Jesus, the University of Rochester physics professor could subject it to the carbon-14 dating method he helped develop—if only Roman Catholic officials would let him do it. For years his hopes gave way to one bureaucratic hurdle after another. But when officials agreed to the test a year ago, he was optimistic once more. Then last month another blow. Three laboratories are to have a piece of the shroud but Gove's Nuclear Structure Research Laboratory is not among them. Gove still hopes to persuade Church officials that three is not enough, that one bad result could make the entire test inconclusive. After so many years spent thinking about the image of that bearded man on the shroud, he deserves to prevail."

It had been written by my friend Lee Krenis, editorial page editor of the D&C and I was very touched by it. Lee had been in the university's public relations department when we first got involved with the shroud and she had always been intrigued by the story. I sent a copy to Donahue, Woelfli, Hall and Canuto, with a little note: "Does this not bring tears to your eyes. I would enclose a small piece of linen to dab them, but you know how it is—Harry."

On 25 November I called Canuto. Rinaldi had told him that Dinegar had a copy of the letter that the three labs sent to the cardinal. No one knew how he obtained it. Rinaldi said a recent STURP meeting held in Rye, New York and attended by Gonella was a total bust. It broke up early. D'Muhala and the rest of the members of STURP were very angry with Gonella. Dinegar asked Gonella if the three labs don't accept the invitation to date the shroud, what happens then? Gonella said he had an alternative plan. He would be prepared to go outside the seven labs.

I then called Rinaldi. He said his position was extremely delicate. He did think that the choice of only three labs was a mistake, but on the other hand Gonella did not. Gonella said that he had an alternative if the three labs turned the offer down—he could go to other labs. I asked Rinaldi which ones he thought he could go to? The heads of all the other labs were close colleagues and I just did not think that Gonella could find any of them who would be willing to take this job on if the chosen three withdrew. I urged Rinaldi to phone the cardinal and he said he was considering doing that. He wanted to wait until he was sure that the cardinal had received the letter from the three designated laboratories. It is clear in retrospect, that a priest does not casually phone an archbishop to give him advice.

On 1 December I talked to Harbottle. He said that he had telephoned Dinegar at the STURP meeting at Rye on Saturday 21 November. Dinegar said there was no point in Harbottle driving to Rye for the meeting. Gonella had dug his heels in with respect to both the question of carbon dating and further tests by STURP. Apparently Turin had decided that even if the three labs agreed to participate and samples were removed this winter, that would not be followed by STURP tests. STURP had had high hopes of carrying out these measurements and were ready to go, but now their hopes were dashed and they were very disappointed. This, undoubtedly, was why D'Muhala was angry with Gonella.

Dinegar told him that he had a copy of the letter that the labs were sending to Ballestrero. He said that he had gotten it from a very unlikely source. Harbottle said that he assumed it was me. I assured him that I had had no communication whatsoever with Dinegar since the Turin workshop and, in any case, I was the least likely person to further foster Dinegar's illusions. He said that Dinegar had this letter in his pocket when he asked Gonella what Turin would do if the three labs turned Turin down. Gonella responded that that was a very unlikely possibility. Dinegar pressed him and Gonella said that, in such a case, he had a contingency plan. He did not say what it was. Harbottle thought that it might be to invite Hall to go it alone, but he doubted if Hall would, as much as he might want to. I mentioned other labs that Gonella might try. There were several; Toronto, Israel, Tokyo, McMaster and Utrecht, but I doubted whether any of them would accept the job, but then again who really knows?

Harbottle said STURP had turned on him. He had received a letter from D'Muhala with a copy of the interview that Harbottle had given to *Science News* in connection with the American Chemical Society meeting in Denver in which he had been described as a STURP spokesman. D'Muhala pointed out that Harbottle had no right to speak for STURP. He was to cease and desist in the future. The letter had been copied to Ballestrero, Gonella and a firm of lawyers. Harbottle was furious. He told Dinegar about it and said that they should kick D'Muhala out. Harbottle reminded me that D'Muhala landed back in the US after the 1978 STURP tests saying that he had proven that the Turin Shroud was indeed Christ's shroud.

On 3 December Peter Rinaldi called. He started as usual by enjoining me not to reveal anything he told me for fear of his being excommunicated. He said that he had spoken to Gonella the previous day. Gonella had said that the cardinal now had the letter from the three labs, but he did not reveal either his or the cardinal's reaction to it. Rinaldi surmised that the cardinal would contact the Vatican. He felt that the Vatican was undoubtedly concerned about their image and might reconsider. Rinaldi had kept emphasizing to Gonella that there would be bad publicity if Turin kept making bad decisions. He had told Gonella about the reports that appeared in the Tucson and Rochester newspapers, including the editorial that appeared in the D&C lamenting that I had not received any sample. He said that the article was really quite good, it reflected well on me and also on the person who wrote it.

On 7 December Canuto told me that Gonella had called Rinaldi to say that he was not concerned about the news reports in the Rochester and Tucson papers—these were just provincial newspapers. He had convinced the cardinal to stick by his decision for three laboratories only.

I phoned Donahue and told him the cardinal had received the letter and that Gonella had advised him to stick with his original decision. Doug was very disappointed. He speculated that if the three labs refused to do the dating, the cardinal might ask Hall at Oxford, Van der Borg at Utrecht, and Erle Nelson at McMaster. Then he remembered that he couldn't ask Hall. I told him about Dinegar claiming to have a copy of the letter that was sent to Ballestrero and asked him if Damon by any chance could have given it to him. Doug said absolutely not.

Neil Cameron called to say that Hall had made an offer to some BBC person (who was not connected with the Cameron *Timewatch* programme) to give the BBC an exclusive on the Oxford dating of the shroud for a fancy price.

In early December 1987, I received a paper from D Allan Bromley who was at that time Henry Ford II Professor of Physics at Yale University. He later became the chief science advisor to President Bush until the latter's defeat in the 1992 elections. It was titled 'The Nucleus, its Impact on Science and Society'. It was prepared for presentation at a luncheon for members of congress hosted by the Division of Nuclear Physics of the American Physical Society, held in the Rayburn House Office

Building in Washington. Part of the talk covered the field of ultrasensitive mass spectrometry using some information he had requested I send him. He ended by saying "Perhaps the greatest immediate interest in this field centres on a worldwide project just now underway to once and for all determine the age of perhaps the most famous relic in Christendom, the Shroud of Turin. Within the next year, seven laboratories around the world under the aegis of the Papal Academy will measure the age of samples taken from the shroud at the same time and under the same conditions as a variety of control samples prepared by the British Museum from fabrics of known age." This was a very nice plug for AMS and for its application to its most newsworthy project up to that time. When Bromley made his presentation he was not aware that the number of laboratories had been reduced to three.

On 28 December 1987 I again talked to Rinaldi. He said that ten days earlier the cardinal had written to the three labs hoping to convince them to agree that there was no chance of returning to the original protocol.

I called Doug Donahue who said he had just received the letter from Ballestrero. Ballestrero (aka Gonella) stated that the reason the labs had given for changing the number from three back to seven was that it would provide a higher accuracy to the measurement, but that was not valid. If three labs cannot give a definitive result, what did this mean for their ability to date other artifacts? Doug pointed out that the letter confused statistical and non-statistical errors (a mistake one would not have expected from a professor of metrology which Gonella claimed to be). He said that he had talked to Hall who had received the letter the previous week. Hall was going to agree to do it. Hall said that Woelfli would also. Donahue had not yet spoken to Woelfli. Donahue said the three labs would probably decide to do it.

So despite all the high-minded statements he, Damon, Woelfli and even Hall had made to me in writing that they would stick by the protocol, it all went down the drain as soon as their bluff was called by Gonella. That was all it was—pure and simple bluff. I was not surprised that Hall took this position but I was deeply disappointed that Damon, Donahue and Woelfli, whom I regarded as personal friends as well as colleagues, did so as well. Would I have done the same if I were in their shoes? I thought not, but one could never know unless one really were put in that position.

Canuto called and said he had tried to persuade Donahue to stand firm. He asked whether the cardinal's letter had really answered the joint letter they had sent to him and Donahue said it had not. Canuto asked then why knuckle under now? That would make the whole workshop a mockery and furthermore the Pontifical Academy of Sciences had been kicked in the stomach. Somehow these arguments did not move Donahue.

I said I thought we should have a press conference and that I would call Harbottle and suggest it. I did so that same day. I told him about the cardinal's reply. I suggested a press conference and he agreed. I said that it should be held before the three labs actually agreed to date the shroud and said that I would call Woelfli tomorrow.

I finally reached him at his home on 30 December. He said that Turin clearly wanted to play the leading role—there was animosity between Rome and Turin as to who controlled the shroud. He said that he would probably say yes, but with restrictions. Woelfli told me that Hall had spoken to Gonella and had suggested that Tite and the representatives from Oxford, Arizona and Zurich should meet with Gonella in London to discuss final conditions. Woelfli felt that if the three did not do it, others would. I asked who? He said Otlet, Harbottle or Duplessy. I was surprised he didn't include Rochester. He also said that the letter claimed that Tite had made no objection to the suggestion that only three labs be involved. I suspect that Tite was merely dancing to Teddy's tune.

That same day, Donahue told me he had also spoken with Woelfli and had been very surprised that Woelfli would now go along. It was because he was afraid, if he did not, Gonella would find a third lab to replace Zurich. Donahue said that he would go to a meeting in London. He would still like to have me involved in the measurement at Arizona, but he would not mention it. I suppose it was because he feared it would infuriate Gonella.

Rinaldi told me in early January that the meeting with Gonella, the three laboratories and the British Museum was to take place in London on 22 January 1988. I told him that we were going to hold a press conference in the very near future. He thought that could be quite influential because it would make it clear to the general public that some responsible scientists thought the project would suffer if only three labs were involved. He said that the articles that appeared in the Rochester papers had caused ripples in Turin.

Harbottle informed me that he could get the use of a room at the Columbia University Faculty Club for the press conference. He suggested that we contact Senator Al D'Amato and/or Daniel

Patrick Moynihan, who are the two senators from New York State. We could note that we are the two New York laboratories who had made the first proposal to date the shroud. We were the developers of both the AMS and the small-counter technique and inexplicably we have been excluded from the dating endeavour. Could the senators by inquiry through our Ambassador to the Holy See find out why two such distinguished laboratories were summarily excluded? He said he thought that this might have the effect of slowing Gonella's headlong rush to get the dating done by the three labs he had chosen.

Rinaldi had told me that Ian Wilson, author of the most authoritative and, in some ways, the most fanciful, book on the Turin Shroud, was opposed to the use of only three labs and he might have some influence with Gonella, so I phoned Wilson on Monday 11 January 1988. He said he had spoken to Gonella and had raised the question of there being no representatives of the small-counter labs involved in the dating. He tried to persuade Luigi to include the small-counter labs—specifically Otlet's. Gonella told him that no more than three samples could be taken from the shroud. Wilson said that Hall was certainly going to agree to do it. The publicity he would receive from dating the shroud would be too tempting for Hall to resist. I told Wilson that we were going to hold a press conference by the week's end. He said that a coordinated effort would be good and that we should include the London *Times*. He made rather negative comments about Chagas, based on what he had heard from Gonella but said that he had heard reasonable things from other people. I assured him that Chagas was a wise and distinguished scientist and a gentleman and that I deeply deplored his elimination from the enterprise. He had done nothing to merit Gonella's malice.

I called Canuto that same day and said that certainly 22 January would be a victory for Gonella. Canuto thought that we should try to meet with D'Amato and persuade him to talk to Ambassador Joseph Kennedy, who had replaced William Wilson as the US Ambassador to the Vatican. Kennedy would be able to speak with Casaroli. He said that in his view, the focus of the press conference that we were planning was that the pope was being given bad advice.

On 11 January I phoned D'Amato's office and spoke to a Jack Kinnecut, a member of D'Amato's staff. He said that he would try to get the name of the appropriate D'Amato aid for science or religion.

I talked to Canuto later that day and asked him if he would try to contact some reporter from *La Stampa* in Turin and also from the *New York Times*. He gave me the telephone number of John Noble Wilford, the current head of the science section of the *New York Times*. Canuto phoned me back that same day and said that the *La Stampa* representative in the USA had phoned him and he had given him details of the press conference. I called the *New York Times* and discovered that Wilford was on assignment in Houston. I gave both my home and work phone numbers for him or some other *New York Times* science reporter to return my call.

On 12 January Harbottle and I talked over the details of our press conference. He said that he had arranged for lunch for noon on 15 January at the Columbia Faculty Club. The people at the lunch would include Canuto, myself, my colleague Shirley Brignall and also Harbottle and his wife.

I made another attempt to contact Senator D'Amato. He was out of town but one of his administrative assistants said that he thought that it was something D'Amato might like to look into. Why were the two laboratories in New York State that invented the new methods of carbon dating and originally offered to date the shroud not chosen to do it? Why not indeed?

On 12 January Canuto called me and said that *La Stampa* had published a notice of the press conference and that Gonella was furious about it. He suggested that I also invite Rinaldi to the press conference and the lunch.

The next day, I called Harbottle. He said that he had just talked to Otlet and Otlet thought that we should pull out all the stops. He said that the cardinal would not give Gonella permission to go to London unless all three of the labs agreed, in advance, that they would date the shroud.

I received a copy of the article that appeared in *La Stampa* on the 13th of January as translated by Rinaldi. The headline read 'Open Letter to Pope Accuses Cardinal Ballestrero—shroud research contested in the USA'. It noted that:

"Two of the most prestigious American laboratories, Brookhaven's and Rochester's, do not accept their exclusion from the research programme that aims to date the Turin Shroud. The archbishop and his research coordinator assert that 'no official decisions have as yet been made'. ...With an 'open letter' to the pope they will protest the recent decision of the Holy See and the Archbishop of Turin, Cardinal Ballestrero to exclude them from the research.

"The statement of the scientists of the two American laboratories will be made known at the Faculty House of Columbia University on Friday 15 January. Their open letter to the pope, owner of the shroud, will make the point that he is being 'badly advised' and that he is making a mistake if he approves a limited or reduced version of the research whose outcome will be, to say the least, questionable. They ask the pope to reconsider his decision and permit them to analyze fragments of the shroud via the carbon-14 test that can date the shroud. ...

"How did the Turin archdiocese curia react to the protest of the American scientists? Said Cardinal Ballestrero 'It seems to me the US scientists are acting as if the final decision had already been made. We are still discussing the situation and will in due time communicate the results.' Rather cautious too is the reaction of Professor Luigi Gonella of the Turin Polytechnic Institute and supervisor of the 'Turin Shroud Research Project, Inc.' a group of international experts who have been researching the Turin Shroud for the last ten years. Says Gonella, 'The Brookhaven and Rochester scientists have no reason to protest. There was no firm agreement with them. If anything it was just a proposal. But aside from this, using a press conference to broadcast their protest is certainly not a laudable procedure. We hope that the serious business of researching the Turin Shroud will not end by becoming a race of who can get there first'."

I received a message from the *New York Times* that my call had been referred to another science writer, Walter Sullivan. He was their most senior and prestigious science writer. He called me that day and I spent about 15 minutes giving him the story. He said he would not be at the press conference, but he now had enough to use and would write it up tomorrow and that it might be in the Saturday paper.

I considered it important to contact Donahue and Damon before we went to New York and so I called Doug on 14 January. Gonella had phoned him requesting a letter agreeing to date the shroud. He and Paul Damon were actually working on that letter now. He said that the letter would say that they were going to the 22 January meeting in London to establish conditions under which they would be willing to date the shroud, but they would not agree to do it unconditionally in advance.

That evening, Damon returned my call. He said he would be in London on Wednesday 20 January and that Donahue would arrive the next day. He said that they would make demands of

Gonella that would make the whole affair workable. I, of course, had continued to hope that perhaps at least Arizona would decide not to go along with Gonella's dictate but it was pretty clear from talking to Damon that they were going to proceed. I knew Damon to be a person of considerable rectitude and decency. Apparently the lure of dating the Turin Shroud was so great it overcame his previously expressed reservations. He had said he was opposed to limiting the number of laboratories to be involved to three, but he now seemed neither remorseful nor contrite about changing his mind. I was surprised and saddened.

Shirley and I went to New York on a flight that left in the early morning of 15th January. We would return that same evening. We arrived at the Faculty Club of Columbia University in time for the luncheon. Present at the lunch were Harbottle and his wife, Shirley and me, Canuto, and Rinaldi.

We then went to the room where the press conference was to be held. I suppose a dozen people or so showed up. Both Harbottle and I made statements about our concern with the revised dating protocol and then we answered questions. Although the turnout was not large, the actual coverage by the world's press was quite substantial.

On 16 January 1988, the day after the press conference, the article by Sullivan appeared in the *New York Times*. It was not up to his usual standards. He emphasized the fact that seven, rather than three, laboratories carrying out the measurement would result in a more accurate measurement. He made the same mistake as Ballestrero, i.e. Gonella, in the latter's letter to the three labs. The real point however, missed by both Sullivan and Gonella, was that a larger number than three labs being involved made it possible to identify a 'rogue' result, if one occurred, as happened in the case of the interlab comparison tests. Seven labs versus three merely increases the statistical accuracy, at best, by 50%.

On Sunday 17 January the Chicago *Tribune* carried a very good article on the shroud. It was entitled 'Shroud of Turin Controversy Resumes'. I was sent a copy of it by Kenneth R Clark, who had authored the article, and also by Tom Beasley, a friend of mine in the Chicago area. Beasley sent me a note reading, "Thought you might like this for your archives (if you are keeping them). It came from the Chicago *Tribune*. Now then, don't get steamed, life is too short. The Holy See has never been particularly ecumenical in these sorts of things. They will come around. Cheers, Tom."

The article by Clark contained the following: "The broadside fired by Drs Harry Gove, head of the physics and astronomy department at the University of Rochester, and Garman Harbottle, his counterpart at the Brookhaven Institute, opened a scene charged with implications of intrigue, political manoeuvering, broken promises, questioned motives and bruised egos that threatened to renew rather than settle the controversy over the authenticity of the shroud, called by the late Paul VI 'the most important relic in the history of Christianity'."

I was quoted as saying that there was speculation that the motive for reducing the labs from seven to three was to conserve the amount of material that would have to be destroyed—that argument was specious. "As a homely example, let's measure it in terms of postage stamps," he [Gove] said. "The entire shroud could be covered by 8800 standard stamps. The amount of material the original seven labs would need is a sample the size of about $2\frac{1}{2}$ stamps. Three labs would need 7/10 of a stamp. So what! It's like knocking a few dollars off the national debt."

I was further quoted in the article as saying that we were particularly annoyed by the fact that the two labs in New York State—the ones that originally offered to date the shroud and the ones that invented the technique that would make the dating possible—were two of the four labs that were excluded. The reason for that was not at all clear to us. We felt that the cardinal was being given extraordinarily bad advice by his science advisor, Luigi Gonella.

Gonella was contacted in Turin and he refused to give the reasons for the reduction in the number of labs. He added that he would not budge in the decision just because Gove and Harbottle were upset about it. He said he did not have to account to Gove and Harbottle but only to his faculty at the Turin Polytechnic. He added that holding press conferences was just an effort to intimidate people who disagree with you. They can hold all the press conferences they want and it will not change anything.

The article said that Dr Paul Damon who, with his University of Arizona colleague Douglas Donahue, had won the coveted right to test the age of the shroud, said the process was extremely complex—it was a daunting process, but they could do it. He said he would like to do it with seven laboratories rather than three for a number of reasons. One of them was a very personal one. "If we go ahead, some of our colleagues, whom we trust and respect, are going to be angry at us. They'll say we should have held out for the original protocol." That was, indeed, precisely how I felt.

On 18 January 1988, Bob Otlet called. He said that he had heard from Harbottle that the press conference had gone well. He said Reuters had carried the story, there would be an article in the London *Times* on Saturday and it was also on the BBC news. Otlet said that Teddy Hall and Sir David Wilson (Wilson is director of the British Museum) were members of the millionaire's club so that one had to be very careful in dealing with them. (He may have to be but I do not.) He said he was still worried about the possibility of collusion between the British Museum and Oxford.

On 18 January Canuto said that Rinaldi had written to Cardinal Ballestrero. It was a very gentle letter—more or less a run down of what happened at the press conference. Canuto said that there were some good articles that appeared in various Italian newspapers—mainly as a result (he said) of his own inputs. A reporter in New York representing the *Journal*, a Rome newspaper, had seen the article in the *New York Times* and sent his paper a summary of it. This reporter then talked to Canuto and, as a result, he submitted a much longer and really excellent article to the *Journal*.

On 18 January 1988 members of the executive council of the Committee for Scientific Investigations of Claims of the Paranormal sent a telegram to the seven labs. They expressed strong support for following the original protocol for carbon dating the shroud.

On 19 January Otlet sent me a fax message which included a brief interview that he had given on BBC Radio 4. They asked his opinion of the changes that had been made in the protocol. He ably presented all the reasons opposing them that we had given in our news release. He also sent me the London *Times* of Saturday 16 January containing an article triggered by our press conference. The headline read 'New Dispute on Dating Tests'. It was an interview with Otlet and it mentioned the fact that the University of Rochester and Brookhaven were very much concerned about these changes in the protocol. There was also an article in the Oxford *Mail*, headlined 'Unholy Row as Church Drops Shroud Test Labs'. Finally he included a UPI dispatch that was also triggered by the press conference. It appeared that our press conference got very good coverage in Britain.

I then talked to Woelfli. He said that at the 22 January meeting in London, they would ask for very restrictive conditions. He said, in particular, he wanted a larger sample than previously requested—the equivalent of 40 milligrams of carbon. He would also insist that there be two independent persons attending the sample taking. He would give these as firm conditions and if Gonella would not accept them then he would withdraw. He said that he had had quite a long conversation with Gonella. Gonella had asked why the shroud was any different from any other important carbon dating Woelfli did every day? I assume Woelfli was patient in his explanation.

Otlet sent me another fax message, dated 21 January with another article from the Oxford *Mail*. This one was dated Wednesday 20 January and was headlined 'Labs Wrapped in Shroud Row'. He included another article from *The Herald* which read 'Row Over Vatican Decision, Harwell Denied Chance to Test Turin Shroud'.

The most interesting article of all appeared in the *New Scientist* of 21 January 1988. It featured a cartoon of two priests standing outside the Vatican. They were holding a piece of paper which read 'Turin Shroud Revised Test Protocol' and one of them, his hands clasped in prayer, was saying to the other, *"It's a miracle, the double blind shall be made to see."* The report was based on a number of interviews including one that I had given their reporter in California. It very much supported the stand that Harbottle and I were taking. It discussed the fact that eliminating both Harwell and Brookhaven meant that the test would be using only one technique and it quoted Teddy Hall as follows: *"The use of two techniques would have strengthened the result. Harwell would have been a good addition."*

On 21 January Sox told me Teddy Hall had gone to the producers of another BBC programme called *The Chronicle*. He had asked them to pay his lab for the story of Oxford carrying out the shroud dating. Neil Cameron's boss was also connected with this programme and he told Hall that the *Timewatch* programme intended to cover the story. He said that *Timewatch* had been on to it a long time before and so it would not make any sense for it to be covered by another BBC programme. Sox claimed that Hall's stock in the BBC was now absolute zero. Cameron would like to cover the dating but if he did so he would like to do it at Arizona and certainly not at Oxford. That amused me because Oxford is an easy drive from the BBC headquarters in London.

I received a call from a reporter named Roger Highfield of the British newspaper, the *Daily Telegraph*, on 22 January. He told me that the meeting had been held. The decision was to use only the three labs, the shroud sampling would be videotaped, they would try to respect the agreement of the Turin workshop as far as possible and the results should be available by the end of the year.

That evening, Harbottle told me he had received a long message from Otlet that verified Highfield's information. The sample taking would be videotaped and a textile expert would be present. He did not know whether it would be Madame Flury-Lemberg. Harbottle agreed that we should pursue D'Amato with great vigour.

On 23 January Canuto said Peter Rinaldi had again talked to Ian Wilson and Wilson stated he was going to write a story deploring the situation. On 17 February Canuto would go to Rome and he would take all the news clips along with him. I should call the *La Stampa* editor in Washington, a man named Corretto, and give him all the facts. Rinaldi claimed that Gonella had told Wilson the samples would be taken at Easter. If there were problems with the carbon dating, more samples could be taken when STURP made its tests in June. This was the first I had heard that the STURP tests were back on track. Finally he said that I should ask Viki Weisskopf, a member of the Pontifical Academy of Sciences and a long-time friend of mine, to intervene in some way or other.

I decided to speak to Rinaldi directly. He said the press release stated that the new procedure still had to be approved by the cardinal. Maybe, if he was sufficiently pressured, he would compromise. There seemed little chance of that to me. Mid-April was the time for taking the samples. Ian Wilson had read him the press release issued at the end of the meeting with Gonella. Rinaldi would try to get me a copy as soon as possible. He was really worried that this whole affair might discredit the church. He confirmed that Gonella had told Wilson that a remedy for a mistake by the three would be more samples taken during the STURP tests. Wilson told Rinaldi he had never seen Gonella in such a black mood, so dictatorial!

On 24 January I phoned Donahue at his home and his wife said he would be home late that night and he would call me tomorrow.

With some reluctance, I phoned Weisskopf at his home. He is one of the elder statesmen in nuclear and particle physics—a giant in both fields and a man of great wisdom and compassion. I had known and admired him for many years. I had had several previous discussions with him concerning dating the shroud. In our conversation, he said that he had always been a little bit worried about the involvement of science in the shroud. He, of course, was not a religious man, but thought that people who were got some benefit from it. He did not think this did much harm. An object like the shroud, that many people believed was Christ's burial garment, should perhaps just be left alone. What was there to be gained one way or another? I said that I had to agree with him and I also expressed my reluctance in actually calling him at all to suggest that he might intervene. He said that he personally did not believe that the shroud was Christ's burial garment but that it didn't do any harm for people to believe it was. He said it wasn't terribly important one way or the other. It didn't compare with the question of nuclear war or birth control or other questions in which the Catholic Church was involved. I said there was absolutely no scientific justification for measuring the age of the shroud. The only thing that concerned me was that if it were going to be done it should be done right. He then said, "The highest desire that I would have with respect to the shroud is, first that it be left alone, and secondly, if it has to be investigated scientifically, that you be involved in the operation." We left the matter there. It was ridiculous of me to think he could or would use his influence as a member of the Pontifical Academy of Sciences to change decisions that had now been made, or that it would have made any difference even if he did. It was another case of my clutching at any straw.

Chapter 9

Laboratories Accept, Samples Removed from Shroud

On 25 January 1988 I again phoned Donahue. It was one of the most difficult conversations I have had for some time. I asked how things had gone. He said he supposed it depended on one's viewpoint, but it was going ahead and nothing was going to stop it. The meeting had been attended by people from the three laboratories, Tite and Gonella. There had been a fair amount of enthusiasm on the part of most of the people there. I asked what mood Luigi had been in and he said that he had been quite agreeable (in contrast to Wilson's perceptions). Every request or demand they had made (and he admitted they had not really made any demands) about how the operation should be carried out had seemed reasonable to Gonella. His response to all of them was that he would consult the cardinal. Donahue said, however, there was still a lot of venom in Gonella directed at me but even more so at Chagas. I asked if he understood why Chagas had been a target of Gonella's enmity? He could only speculate that it had looked to Gonella as if Chagas was taking over the whole enterprise. He said that it was clear that Gonella and the cardinal are in complete control. I asked him what sort of safeguards they had managed to extract from Gonella. He said he would read me the press release. It had been issued by the British Museum on Friday. There was a reporter there from one of the papers—I asked if it were Highfield, from the *Daily Telegraph*? He said that was right. Highfield had hung around most of Friday looking for news. After the meeting he (Donahue), Tite and Gonella had composed the following press release:

"Representatives of the three radiocarbon dating laboratories, Arizona, Oxford and Zurich, accepted by the Vatican to undertake the radiocarbon dating of the shroud met on 22 January at the British Museum together with Professor Luigi Gonella, the scientific advisor to the Cardinal of Turin and Dr Michael Tite of the British Museum who had been invited to help in the certification of the operation. After discussion, they accepted the decision of the Vatican to use no more than three samples in the interest of conservation of the shroud. Procedures for taking the samples from the shroud and for the treatment of the results were discussed and proposals on this will be submitted to the Archbishop of Turin for his agreement. It is proposed that, as far as possible, the spirit of the original protocol of the 1986 meeting be retained. Each laboratory will be provided with control samples of known age. The samples will be taken from the main body of the shroud away from patches or charred areas under the supervision of a qualified expert. Certification of the samples will be undertaken by the Archbishop of Turin, Anastasio Cardinal Ballestrero, Pontifical Custodian of the Shroud of Turin and by Michael Tite of the British Museum. Representatives of the three laboratories will be present in Turin to receive the samples. The overall procedure will be fully recorded by video film and photography. The laboratories will submit their results for statistical analysis to the British Museum and to the Institute for Metrology 'G Colonnetti'. The timetable for the operation has not been established but it is hoped that the radiocarbon dates on the Shroud of Turin will be released by the end of 1988. If these proposals are approved by the cardinal, then a letter will be submitted to *Nature* giving further details of the procedure. The participants of this meeting wish to take this opportunity to record their appreciation to Professor Carlos Chagas, President of the Pontifical Academy of Sciences, who chaired the original meeting in Turin in October 1986 as well as the other participants who played a crucial role in moving the project forward."

Donahue said that embodied everything we now knew. I asked whether the textile expert would be Madame Flury-Lemberg? Donahue said it would be some Italian expert—for purely political reasons. At one point they actually suggested Madame Flury-Lemberg and Gonella was adamant in his opposition to her. Donahue said that it was pretty clear from someone—the pope or God knows who—this was the way things were going to be done. It was also clear that the people at STURP were chomping at the bit. Gonella had also made some really nasty comments about Chagas interfering with their measurements. That was where the venom really showed.

I told him that I had just finished talking to Weisskopf, principally because he is a member of the academy, and that he had told me that he felt very badly about the way Chagas had been treated. Weisskopf essentially had advised me to stop fooling around with the shroud. He had told me that he would prefer to see nothing done; however, if something were going to be done, he would really like to have me involved. Donahue said those were his sentiments also.

I added "It will be a lot of fun. You've got to admit, Doug, you are going to have some fun doing it." Donahue said, "I am not going to admit that. I think the height of the fun was the meeting at Turin. I think it's downhill from then on as far as I am concerned. I don't especially enjoy dealing with the press, but on the other hand it gets dealt with in ways I don't like." I said that I should just remark that in all the contacts we had had with the press, the people we had been criticizing were Gonella and STURP, not him or any of the labs involved. I hoped there were no hard feelings as a result. Donahue said there were not.

We discussed the article that appeared in the *New Scientist*— especially the cartoon. Donahue had seen it and thought it appropriate and quite touching. I asked whether they would be prepared to accept unravelled samples and he said no, the samples would not be unravelled. I then said he would know which sample was the shroud. He said no he did not know what it looked like. I offered to send him a photograph. Donahue said he could look it up in the *National Geographic*. He said that Tite would make an effort to make the samples look as similar as possible but there was no pretence that it would be really blind because the shroud weave was so distinctive.

Toward the end I said, "I can't disguise from you the fact that I envy the hell out of you." Donahue said, "I know that and I feel for you, I really do. I said before I am going to do my best to have you here as an observer if you can make it." I said that would please me.

I phoned Canuto the next day and related the gist of the above conversation. He asked if I had criticized Donahue for knuckling under to Gonella. I said no I had not, what good would that do? He said that I should send a letter to the pope. It would be the last resort before the death sentence was carried out. I should do this via both the US ambassador to the Vatican and the Vatican ambassador to the UN and get Otlet and Harbottle to sign it. I said I would think about it.

I phoned Kinnecut, the person I had spoken to before at Senator D'Amato's office, and he said that he had gotten nowhere. However, the message was clear—D'Amato was really not interested in helping one of his constituents—at least not on a matter like this. That was all the more reason for my continuing never to vote for him.

One of the next things I did—another last-gasp effort—was to write a letter to Sir David Wilson, the Director of the British Museum, dated 27 January 1988. I enclosed a copy of the press release issued by the British Museum following the 22 January meeting. I said that I had no reservations whatsoever concerning Dr Tite's honesty, integrity and credibility as a representative of the British Museum in this enterprise. However, there were many people who were overly suspicious of the entire operation. The situation was particularly exacerbated by the fact that the head of one of the three laboratories to be involved, Professor E T Hall of Oxford, was also on the board of directors of the British Museum. I pointed out that the original protocol called for a third person to be involved in both the certification and data analysis, namely the president of the Pontifical Academy of Sciences or his representative. I said that Dr Chagas was such a distinguished scientist that if both he and Dr Tite had been involved and if the original seven labs had participated, the enterprise would have been as credible as possible. I was astonished that Wilson would permit the British Museum to risk having its reputation called into question in what had become a somewhat shoddy enterprise. I was afraid that four of the participants at the meeting, namely Damon, Donahue, Woelfli and Tite had bowed to the dictates of Gonella, warmly supported by Hall. I ended by saying that I feared, sadly, that Mike Tite had taken on a responsibility which he and the British Museum might live to regret.

I received a letter dated 2 February 1988 from Sir David: "Dear Professor Gove: Thank you for your letter of 27 January 1988, I have noted the contents. Yours sincerely, David Wilson." At least he had been civil enough to acknowledge my letter.

David Sox sent me an article he had written for *The Tablet*, 30 January 1988, headlined 'Dating the Shroud'. In it he stated he had found it extraordinary that no one other than he seemed to feel that the shroud may prove to be a fake. He pointed out that there were some formidable obstacles to its authenticity. First, it emerged in history only in the 14th century. Second, no archaeological finds had been made of linen of herringbone weave dating from the time of Christ in Egypt or Palestine. Third, the blood marks appeared to be too precise in detail to have been left on a cloth that enclosed a body taken from the cross to a tomb. Moreover, the stains were still red which would be very strange after all this time. He did concede that the shroud had passed a battery of tests. Its image was a photographic negative and, with the exception of an isolated claim by one scientist, no traces of paint had been discovered. The anatomical detail appeared far too good for a medieval artist. The dating procedure soon to be carried out would be "the archaeological test of the century". Shades of Robert Dinegar!

Sox wrote a note to me saying that his feeling was that, "although it is not perfect, and even far from it, let's get on with the test. Waiting even a year would be waiting another decade. One has to realize that a new archbishop and a new Gonella will be taking their place in Turin and that Ballestrero...has moved very far against great opposition." He said that I would be linked with the Arizona effort. The word he had was that the test results would be available at the end of the year. Although he would have preferred a Rochester and Brookhaven involvement and an involvement of more than three laboratories, he felt that Oxford, Arizona and Zurich could arrive at a date that is clearly medieval or near the time of Christ. He had attended his first British Society for the Turin Shroud meeting in seven years, and said, "No one is ready for the possibility of the shroud being a fake."

On 1 February 1988 I spoke to Canuto. He said it appeared that Chagas had lost power in the Vatican. Chagas was very disappointed that the three labs had bowed to Gonella. Canuto felt that a letter to the pope was very important, if only to give him some historical perspective on what was going on.

I proceeded first to compose a letter to Cardinal Ballestrero that I hoped would be signed by both Harbottle and Otlet as well as me. I also would have liked it to be signed by Duplessy, but had been having great trouble in contacting him. I phoned Corretto, the reporter in Washington for *La Stampa*, and sounded him out on the idea of publishing an open letter to Ballestrero in *La Stampa*. He said that he thought the idea was excellent and he would be happy to do something about it.

On 15 February I faxed both Otlet and Harbottle the open letter to Ballestrero I had composed and a request they indicate to me whether they would sign it. I related what Corretto had told me.

He thought *La Stampa* would welcome such a letter and that it would have some impact. I said I thought it would be good if we could get Jean-Claude Duplessy to sign it as well but had not been able to contact him.

On 26 February Otlet replied. He pointed out that, because his lab was just a part of the main Harwell establishment, everything had to be vetted by their press officer. He said they were not prepared at this stage to support any further press contacts regarding the Turin Shroud. He went on to say that his own private feeling was that this open-letter approach would not do any good. "It will be seen as a stunt and will be used as good evidence against you as the very reason your laboratory and the ones that sign with you were best left out of the dating exercise. Could you not try a direct letter to the pope in your own hand?" He suspected that it would only be through moves at the highest diplomatic levels that would cause people to rescind their plans. He said: "The biggest let-down that we have to bear is that those people whom we counted as our colleagues have not the decency to say that the decision to leave out the founder laboratories was both scientifically and morally wrong." He ended by saying that Duplessy opted out some time ago and would have nothing to do with any publicity. It was now clear that without the signatures of Otlet and Duplessy, the open letter in *La Stampa* to Ballestrero was dead.

On 26 February Donahue told me that there was now going to be 40 milligrams (about 2 square centimetres) of cloth for each of the three labs. He said that he thought the cloth-to-carbon ratio was a factor of eight so this would yield 5 milligrams of carbon for each lab. I said no, it is not, that it was more like a factor of five so they would each get 8 milligrams of carbon. He said that Tite was writing a letter to *Nature* describing the new protocol and in it he would say that there will be 40 milligrams of cloth per lab. The interesting thing is that this decision now provides 120 milligrams or 5.7 square centimetres of cloth to be distributed equally among the three labs as opposed to the $12\frac{1}{2}$ square centimetres that was agreed to in Turin at the workshop. That meant that they would be using almost half the amount of cloth originally agreed to at the workshop. That really makes the argument of conserving cloth specious.

In late February, I again talked to Harbottle. He noted that the other senator in New York State, Daniel Patrick Moynihan, would be up for re-election. We should send a joint letter to him with appropriate newspaper clippings and also mention that my previous contacts with Bishop Clark of Rochester made it clear he was on our side.

On 26 February 1988 I put down these final thoughts on the shroud project, just for my records:

"Far and away the saddest and most deplorable aspects of the whole shroud dating enterprise are, firstly, the fact that a group of estimable people comprising Canuto, Chagas, Damon, Donahue, Duplessy, Hall, Harbottle, Hedges, Otlet, Tite, and Woelfli, most of whom were colleagues and some of whom were personal friends, now have a distinctly different and more suspicious attitude toward one another and secondly, a scientific investigation that could have been exciting and challenging and, above all, a great deal of fun, has turned sour. None of us has come out of it whole and pure although some more than others. It is extraordinary to me that twelve people including me, each one of whom has a vastly greater scientific standing than Professor Gonella, should have allowed such a mean spirited person to call the tune to which all of us danced in one way or another.

"As for me I will write a personal letter to the pope that he will probably not even read and then let the matter rest. If I am invited to be a guest observer of the measurements at one of the three laboratories I will probably accept. Some day I hope to be involved in writing an account of the affair which has now spanned eleven years of my life and entailed considerable effort on my part. A lot of the effort was enormous fun however. All along, my chief motivation was great curiosity as to the shroud's age, the realization that it was a perfect artifact to demonstrate to the general public the power of AMS, and the desire that the measurement be made in the most credible fashion.

"Unfortunately it is not now going to be made in as credible fashion as it could have been and, what is worse, no reason has been given for the change in procedures. Obviously the reasons are so contemptible as to be embarrassing. However, maybe luck will prevail and the job will be done well enough. That, however, will not compensate for the fact that the affection and admiration some of us had for others in the group of twelve has lessened and, in some cases, even vanished. For that, I will never forgive Gonella. Knowing him as I do, I do not expect that will bother him in the least."

A little later that afternoon, Peter Rinaldi called me. He said there had still been no approval by the cardinal in Turin for the decision reached in London at the 22 January meeting. He thought more and more that it seemed as if Gonella had been rethinking his position. He had been in touch with Lukasik. On 27 February Lukasik wrote a letter to Gonella, and Rinaldi said he would send me a copy. Lukasik was beginning to think that three labs were too few and Rinaldi pointed out that Lukasik had great influence on Luigi.

Meanwhile, I was trying to make contact with Daniel Patrick Moynihan, our senior United States senator from the State of New York. I managed to contact one of his staff people, Roy Kienitz. He was the person who covers science matters for Moynihan. I talked to him on 3 March and he said he would talk to Moynihan's chief of staff. Obviously it was a rather delicate matter and he took my phone number and said he would get back to me.

Rinaldi sent me a letter Ian Wilson had written to Rinaldi, Dreisbach, Otterbein, Maloney and Meacham. He said he recognized much of the wisdom of the 1986 protocol, but he had always regarded the idea of providing the shroud samples to seven labs as unnecessarily excessive. He said whether seven or three laboratories happened uniformly to indicate a 14th century date, he would still want unequivocal evidence of the hand of an artist. Conversely, if they happened to indicate uniformly a first-century date, he was sure that those who believed the shroud to be the work of a forger would justifiably want some additional evidence of exactly how the image might have been created.

There was another document written by Wilson. It was titled 'A Small Matter of Date', and the subtitle was 'Fresh perspectives on the row over the dating of the Turin Shroud'.

He said the issues raised by the shroud were extraordinarily clear cut. It was either one of the most artful forgeries of all time, the creation of a medieval mind (for it could be reliably traced at least back to the 14th century) so ingenious that he or she had deceived a whole battery of arguably competent doctors, physicists, chemists, archaeologists and historians, or it was the genuine burial wrapping of someone who had suffered at Roman hands a crucifixion so close in all its details to that of Jesus Christ, that identification with Jesus would seem more than probable.

He noted that, in the public mind, there was no test considered more crucial to determining which of these two alternatives was right than the radiocarbon dating. He approved of applying this ultimate test to the shroud but warned that it was not infallible. Several radiocarbon laboratories had developed sophisticated new techniques whereby the cloth could be dated from mere fingernail size samples and that considerable accuracy could be expected.

He said, however, such passions had been raised by the selection of only three laboratories that the decision was unlikely to pass without challenge. Gove and Harbottle had gathered some highly respected independent experts to support their position that any results achieved by the use of the three chosen laboratories might be insufficiently trustworthy to carry the desired level of public acceptance. They were talking of exerting diplomatic pressures to get the pope to overrule the Cardinal of Turin. They had even considered taking a full page advertisement in the major Italian newspaper, La Stampa. (I wondered where Wilson was getting all this information—I never did find out.)

He noted that much of the heat of the present situation could be reduced if dating by Harwell, no friend of Oxford, could be slipped in as an extra safeguard. Perhaps some informal arrangement between the two labs, that is Harwell and Oxford, might even yet be worked. (Damned unlikely I would say.)

His last paragraph was amusing. "And amidst all this most unholy row, one cannot help recalling from the Gospel of St John how even the Roman soldiers who crucified Jesus, pagans as they were, were at least civilized enough to cast lots to decide between themselves as to who should receive his clothing. Might there not be even now someone looking down on the sorry scene and murmuring 'Father forgive them for they know not what they do'..."

On 17 March 1988, Rinaldi phoned and said the mystery had deepened. There was total silence from Turin. Gonella was speaking in riddles. Great secrecy would surround every move that will be made from now on. STURP would play no role in the operation. Samples would be taken any day now. I told Peter that I was in a state of burned out paralysis.

He phoned again later on in the afternoon. He claimed that Wilson had learned from Gonella that STURP would not be allowed to make measurements until after the shroud is dated. Wilson thought this was a very bad move since he believed that a carbon-14 date must be bolstered by other scientific data. I told Rinaldi that this was absolute nonsense. Other tests should take place only after its age was known since one would surely proceed differently if the age were 2000 years old than if it were much younger.

On 18 March I talked to Ted Litherland. He had just come back from a trip to Oxford and he said that he had had a grand dinner at Teddy Hall's house but Hall seemed a bit ill at ease. He was defending his stand on the whole carbon dating enterprise on the grounds that Tite was involved. There was a four year plan that was prepared for Hall's Oxford lab, that stated Oxford was 'chosen' to date the shroud and this was reason enough to support the lab. Ted said that Hall would be retiring in the next year or so and Tite was the top contender to replace him. That was the first time I had heard of this possibility and it turned out to be true. No wonder Tite made no objection to only three labs being involved as long as one of them was Oxford!

The 24 March 1988 edition of *Nature* contained another letter from Denis Dutton. He expressed the worry that nobody had come forward with procedures to secure the authenticity of the samples. He deplored the reduction of the number of labs to three. Shut out from the tests would be Dr Harry Gove of the University of Rochester and Dr Garman Harbottle of the Brookhaven National Laboratory, as well as the Saclay laboratory of France and the Atomic Energy Research Establishment at Harwell. Of equal importance was the fact that the Vatican officials in charge of the test had still not come forward with procedures to secure the authenticity of the samples—procedures, for example, to make it impossible for ancient mummy linen to be surreptitiously introduced into the chain of evidence.

Dutton is clearly an eminent and respectable man but he was certainly snatching at straws here. I don't think anyone in the seven carbon dating labs ever worried that there might be a substitution of Egyptian linen for the shroud—at least I certainly did not.

On 24 March 1988 I completed the final version of the letter that I was planning to send to the pope. I knew it was a lost cause but I wanted to leave no stone unturned.

In summary, the letter gave the background of my involvement in the enterprise to date the shroud. It included all that had take place between the first AMS measurements at Rochester in 1977 and the meeting of the three carbon dating laboratories with Gonella in London on 22 January 1988. It listed the changes in the Turin Workshop Protocol demanded by Gonella and it appealed to the pope to intercede with Cardinal Ballestrero to revert to the original protocol. Enclosed in the letter were the article I had published in *Nuclear Instruments and Methods in Physics Research* on the workshop protocol, our cable to Ballestrero warning him of the possible dire consequences of changes in the protocol, and the letter from Ballestrero to all the workshop participants ordering the changes.

On 25 March 1988 the letter to the pope with the three enclosures was mailed from the main Rochester post office on Jefferson Road. Colourful postage stamps that included 4 cat stamps, 2 T S Elliott stamps, 2 William Faulkner stamps, and 1 stamp commemorating lace-making in the US were affixed. The clerk at the post office was really intrigued by this and she helped me select the stamps and helped me apply them to the envelope in an artistic manner. They were hand postmarked and sent first-class airmail. I did it this way in the hopes that, with such a strikingly stamped cover, it might actually get to the pope rather than being thrown in a Vatican wastebasket.

On 28 March 1988, I wrote a letter to the Honorable Daniel Patrick Moynihan as follows:

"Dear Senator Moynihan: The attached letter and enclosures were airmailed to His Holiness Pope John Paul II on 25 March 1988. Despite the fact that the Post Office clerk in the central US Mail Office in Rochester fancied up the envelope with all sorts of eye catching stamps I fear it still may not come to His Holiness' attention. Hence this appeal to you.

"I have had several phone conversations with Mr Roy Kienitz in your office concerning the possibility that you might ask the US Ambassador to the Vatican to, in turn, inquire of the Cardinal Secretary of State for the Vatican why the two laboratories in New York State at the University of Rochester and at Brookhaven National Laboratory, were not chosen to participate in the radiocarbon dating of the Turin Shroud. The technique for accommodating small enough samples to permit an essentially non-destructive carbon dating of the shroud was invented at Rochester using a fundamentally new technique involving nuclear accelerators and, somewhat later, at Brookhaven where the standard decay counting technique was pared down in terms of sample size.

"It occurred to me that the best procedure at this late date (samples may be removed around Easter) would be for you, if you were kind enough, to forward this copy of my letter and the enclosures to the U S Ambassador to the Vatican with a request that he bring to the attention of the Vatican Secretary of State the fact that this material had been sent directly to the pope.

"I am sorry to trouble you with such an apparently inconsequential request but I feel that the Archbishop of Turin Cardinal Ballestrero is receiving incredibly bad advice from his science advisor Professor Gonella on the most credible way to date the Turin Shroud. Furthermore, it is unbelievable that the laboratories which invented this new technique not be permitted to be amongst those applying it to such an important artifact. The fact that they are both located in New York State emboldens me to bring the matter to your attention. Yours sincerely, H E Gove, Professor of Physics and Director."

On 30 March I wrote to Donahue that it now appeared certain that Gonella's will had prevailed and the three designated laboratories would date the Turin Shroud. I would therefore respond positively to an invitation to be present when the measurements were made at Arizona. After having played a significant role in getting the shroud to the point of being dated, I would hate to end my involvement without at least actually seeing the AMS technique applied to its most famous sample. Gonella had done his best to make the measurement of the shroud's age a joyless experience but I still thought it would be marvellous fun. If I was not to be allowed to do it here in Rochester, then the next best thing for me would be to see it done at Arizona.

In a letter to *Nature* of 7 April 1988, Michael Tite, as promised, gave the procedures that would be followed in dating the shroud as agreed to at the 22 January meeting in London. There were no surprises. He did state that the samples would be provided whole without being unravelled. He noted that even if the samples were shredded it would still be possible for the laboratories to distinguish the shroud sample from others. That was something I had known all along.

In a letter dated 8 April 1988 Tom Beasley enclosed more newspaper clippings from Chicago newspapers with a short note: "Harry: Are you getting bored with this? TB" One of them was an article in the Chicago *Sun Times* of Friday April 8 headlined 'Modern Technology May Finally Fix the Age of the Shroud of Turin'. The article was by Robert Glass. "I would be rather awed if it looked extremely probable that it was His burial cloth", said Oxford University's Professor Robert Hedges. Dr Tite said in an interview, "If you get a medieval date then clearly it is not Christ's shroud. If you get a date of the period of Christ then that doesn't tell you anything, except that it could be the genuine thing."

Gonella was quoted in the article as saying the church never claimed that the shroud was authentic, only that it might be and must be treated with respect and caution. He added that the archdiocese had never agreed to allow more than three labs to participate. "Critics talk about us reducing the number of labs from seven to three while we see it as increasing it from one to three," he said in an interview.

Dr Hedges, head of the radiocarbon dating unit at Oxford's research laboratory for archaeology and the history of art, said he would not try to deliberately determine which of the three samples was from the shroud. "In a sense it almost puts one under too much pressure", he said in an interview. "It is sort of more relaxing if one doesn't really know. Either way," he added, "it wouldn't affect the findings."

Around the middle of April, I received a copy of *Science News* dated 16 April 1988, sent to me by its editor Patrick Young. It contained an article headlined 'Controversy Builds as the Shroud Tests Near'. It noted that some scientists expressed concern about the three-lab decision. Among them were Harry Gove of the University of Rochester, New York, Physics Department, whose lab developed the accelerator carbon-14 dating technique, and Garman Harbottle of Brookhaven National Laboratory in Upton, New York. Harbottle was quoted as saying there appeared to be about a one in five chance in any given measurement that the answer would be wrong. If there were only three labs he said, it might be difficult to identify whose was the spurious reading.

In the article, geoscientist Paul Damon, co-director of the Arizona test, played down that concern pointing out that they hope to get a number of carbon-14 analyses from the 2 square centimetres of shroud being supplied to Arizona, perhaps as many as 7. Dinegar was quoted as saying that Harbottle and Gove, whose labs did not routinely date archaeological samples, were just unhappy their labs were eliminated because they did not meet the archbishop's requirements. (What requirements—that we kiss the hem of Gonella's surplice?) Dinegar said that although he would like to see a counter lab like Harbottle's included "we can certainly get a viable answer from the accelerator method". (Marvellous!—that from a chemist with no expertise in accelerators or carbon dating. He uses 'we' as if he were still involved.)

I decided to write a letter to the editor of *Nature* in answer to Tite's article and it was published in their 12 May 1988 edition.

I compared the new procedures for dating the shroud as stated in Tite's previous letter with the Turin Workshop Protocol. I concluded with, "All these unnecessary and unexplained changes unilaterally dictated by the Archbishop of Turin will produce an age for the Turin Shroud which will be vastly less credible than that which could have been obtained if the original Turin Workshop Protocol had been followed. Perhaps that is just what the Turin authorities intend."

Around 21st April, Peter Rinaldi phoned. He said that complete secrecy continued to surround the sample removal. The 25 April date had been cancelled and the shroud sampling might take place either before or after. No one except Tite and Gonella plus an Italian textile expert would be involved in actually cutting the shroud. It would be videotaped by the Italians, as previously promised. He told me that Ian Wilson had called Gonella around the 7th of April to see whether there could be some STURP measurements before the carbon date was announced and apparently Gonella said there was no possibility of that. Wilson is sickened, according to Rinaldi, by Gonella's unresponsiveness. STURP would not be involved in anything—either the sample taking or any tests before the carbon-14 results were announced.

On 25 April at 11 am, Harbottle called. He had learned from Otlet that the shroud samples had been removed on 21 April 1988. Hall had flown into London on 25 April with the samples in hand and he received a lot of publicity. The archbishop had been, according to Harbottle, furious about Hall's trying to commercially capitalize on the venture. Harbottle also said that the BBC were going to film the measurements at Zurich. He said that, according to Otlet, there was no possibility this time of any outliers because the three labs would consult together so the answers would come out the same. I must say I thought that Otlet was being either paranoid or surprisingly cynical.

I then talked to Donahue. He said that he had been in Turin for the sample taking. The Chagas badge had not been picked up. They had been summoned at 6:30 in the morning for the sample removal. I phoned Damon that evening and he said that he was going to tell Gonella that I was to be invited to Arizona to watch the dating (I wondered why he felt obliged to inform Gonella). He said that he was not going to ask him whether I could come, he was simply going to tell him that I was coming. He said that Oxford might have a problem. They had recently switched to using carbon dioxide gas in their ion source but the system had not been fully tested. They would probably have to convert back to their former ion source before they could do anything about the shroud and that would take some time.

I wrote to Damon on 26 April saying that he had a perfect right to invite anyone he wished to view the measurements at Arizona. It was quite unnecessary to inform Gonella he was inviting me and, in fact, might lead to complications. However, it was his business. I continued:

"As you know, I have been highly critical of the many changes that have been made in the protocol agreed to at the Turin workshop. That procedure would have been the most credible way of dating the Turin Shroud. I think all of us agree that the present procedures are less than ideal. My main criticism is that three labs are too few. I have always made it clear that the three labs that were chosen are excellent ones. If none of the three makes any mistake in their measurements then all will agree with each other within a standard deviation or so, on all three samples. If the public believes there has been no collusion, then the date for the shroud will probably be generally accepted except by those who have emotionally fixed ideas about its age and who would not accept any date that disagreed."

On 26 April Bob Otlet sent me a newspaper clipping from the London *Independent* of Monday, 25 April. The headline read 'Analyzing the Strands of Time' by a reporter named Nicholas Schoon. It said he had met the professor whose delicate task it was to try to date the Turin Shroud, and it contained many revealing stories about Hall. It said the man on the flight from Turin to Heathrow had something very special in his briefcase—a small steel vessel containing a 2 square centimetre piece of linen. This snippet was cut from the Turin Shroud. The article continued:

"Professor Edward Hall, head of the Oxford University Research Laboratory for Archaeology and the History of Art, is charged with the task of determining whether the Turin Shroud dates back to the first century or is an extremely clever medieval forgery. He has just collected his sample from Italy along with 2 other pieces of ancient linen. He has no way of telling which is which, they are simply numbered 1, 2 and 3. [Since the samples were not unravelled it would be instantly apparent to Hall which one came from the shroud—as he well knew. Hall continued to play the 'blind measurement' game.]

"Hall, 63, describes himself as 'a total agnostic' and admits he was highly sceptical about the shroud's authenticity at first. 'I thought it was a load of codswallop, definitely a forgery. But now I have looked at the evidence more closely and have a more open mind. If it turns out to be from the year zero and one had no idea how the hell it was made, I would find that very worrying, quite strange, really'.

"Hall...is soon to retire and his dating of the shroud should crown an unusual and perhaps anachronistic career in science. He is one of the last gentlemen scientists, blessed with a large enough inheritance from his grandfather to pay his own salary. He is quite unembarrassed by his wealth and entirely content that he has found useful and fascinating ways of spending it. He has the world's largest private collection of ancient Chinese porcelain and gives three addresses in Who's Who.

"One of his main preoccupations now is raising 750 000 pounds to fund his professor's chair in perpetuity when he retires. He says, 'I think I will succeed. I've put in 10% myself and luckily I've got some rich friends.' He also hopes a Sunday newspaper will pay a large sum for the rights to the shroud dating story.

"Hall heads what is probably the world's foremost laboratory for applying science to the problem of history and archaeology. It all began in 1953 when he was a postgraduate student at Oxford's Clarendon Laboratory under his professor Lord Cherwell. [Cherwell was Winston Churchill's chief science advisor during WW II.] The young Eton-educated Hall was building a mass spectrometer... Lord Cherwell wanted to improve the image of scientists. He was weary of their being regarded as philistines, only building atom bombs. One day he said 'Hall, it is time to make an honest woman of Science.' He wanted a marriage of the two polarized cultures of art and science, so the mass spectrometer was put on one side and Hall...turned his attention to non-destructive methods of testing ancient relics.

"Soon he had his big break, helping to expose Piltdown man as a fake. In 1912 an apelike jawbone with fragments of a human like skull was discovered on a Sussex heath. It subsequently turned out to be a modern human skull with the jaw of an orangutan with filed down teeth attached. But at the time it was hailed as a vital missing link, and had pride of place in the Natural History Museum for decades. In the early 1950s several eminent anthropologists and zoologists doubted its authenticity. Hall was able to show that the skull bones had been stained deep brown by boiling in potassium dichromate to make them look as if they had lain in the dark soil for thousands of years. His technique was to beam X-rays on the skull and then analyse the wavelength of the X-rays given out by the atoms of stain on the surface. He recalls waiting with scientists from the anthropology and zoology departments while the apparatus ticked over. 'I sent out for lobsters for lunch as one did in those days,' he said. 'Piltdown man was a very good piece of luck and got me and my work a good deal of publicity.' As a result, Lord Cherwell won financial backing to found the new research laboratory. 'There was no nonsense about advertising for the post of director', chuckles Hall. 'He said you'll do it won't you Hall and that was that'.

"Ten years ago, to date an object, a large sample of it had to be destroyed in order to count enough carbon-14 atoms. This was clearly unacceptable for precious relics like the shroud. Since then the high-energy ion accelerator has been developed for carbon-14 dating and has reduced the sample size needed by a factor of 1000. The first demonstrations which showed that this new technique could work were done in the US and Canada. 'As soon as we heard of these developments we abandoned our own approach to the problem and started getting money for the new kind of machine.' The machines cost more than 500 000 pounds. The one at Oxford is funded by the Science and Engineering Research Council as a national research facility.

"The Church's decision earlier this year that only 3 laboratories should date the shroud has caused great controversy. Four others which had hoped to be involved have been excluded, including 2 which had helped to develop the new technique. Now they have suggested that the outcome will be open to criticism. 'I'd be hopping mad if I wasn't chosen,' said Hall. 'But having only 3 labs does not undermine the validity of the dating. I think it was absolutely the right decision. You only need one lab to get it badly wrong to confuse everybody and the chances of that are higher with 7 than with 3'."

Hall had conveniently forgotten that in the British Museum interlaboratory tests, it was only because six laboratories had been involved that it had been possible to identify the one outlier measurement. I found the article quite amusing. It was just the kind of publicity that Teddy revelled in.

On 26 April I talked to Neil Cameron. He said that a week earlier they had had in mind completing their shroud programme for October, but now they had pushed it up to June, so they needed to get the programme taped as quickly as possible. He said that Gonella expected that people would talk to each other at

Dubrovnik and he wanted to do a BBC programme beforehand, otherwise it might not be newsworthy. He said Giovanni Riggi was the one who cut the samples and had also organized the videotaping of the sample taking. The BBC eventually paid Riggi to have the rights to incorporate this video tape into their programme on the shroud. He wanted me to come over to London to be involved in the programme. He was planning to go to Zurich to film the actual dating. I made reservations to fly to London on 12 May and come back on 16 May. I was hoping I would recover the cost of the ticket from the BBC. On the 28th of April, I phoned Sox and told him of my plans.

Cameron called me the next day. He said that he wanted to do the BBC filming on Saturday morning 14 May. He said the BBC would provide $500 plus anything else he could scrape up. He claimed that Woelfli was concerned that STURP might have switched the samples. (That was impossible—STURP played no role in the sample taking.) Woelfli said he didn't think his results would be ready to go to the British Museum until just before the meeting in Dubrovnik. Cameron said he could have sent a BBC crew from New York or asked me to go to New York, but he would rather do the interview with me himself in London. He told me that Hall was not going to Dubrovnik.

On 29 April Donahue called. He said their machine was sparking at voltages lower than it should, so it meant that it would have to be opened and the diodes replaced. They would try to run at 1.7 million volts through next week and if it worked they would run the first shroud sample on Friday 6 May. If the accelerator refused to perform properly they would postpone the run until 23 to 27 May and would open the machine. I should purchase an air ticket and phone Wednesday 4 May to see whether the run was still on. If the run of 6 May goes through, they still plan a second run 23 to 27 May. The reason for these two times was that they were the only times Damon would be there in the immediate future. Doug said Damon still claimed he was going to tell Gonella I was invited as an observer.

On 5 May Brian Kahle from Channel 7 in Buffalo phoned. He had invited me to appear on 17 May on a television programme he hosted from 10 am to 11 am on Channel 7. The other two people on the programme would be Joe Nickell and Benjamin Wiech—a lawyer in Buffalo. Wiech and I had corresponded and talked to each other on the phone many times in the past. Nickell was a passionate disbeliever in the shroud and Wiech an equally passionate believer. I agreed to do this even though by the time I went on the programme, I would probably know what the Arizona carbon-14 results were.

Chapter 10

Shroud Dated, Public Rumours of a 14th Century Date

Shirley and I arrived at the Doubletree Hotel in Tucson, Arizona at about 4 pm, Thursday 5 May 1988. Both of us were keyed up for the event that lay ahead. I phoned Donahue at the University of Arizona's AMS laboratory. He suggested I come to the Physics Department at 8 am the next morning. They would begin the preliminary preparation for their first run on the shroud at 7 am. The actual measurements should start about an hour later.

Around 5 pm, Paul Damon called and suggested he come to the Doubletree and have a beer and a chat with us. He arrived half an hour later. He said when he and Donahue had returned from Turin on the Saturday after the sample had been removed from the shroud, they had decided to divide the total area of about 2 square centimetres (0.3 square inches) they had received in Turin into four pieces each about 0.5 square centimetres or 1/4″ × 1/4″ in area and to store them in different places. One, that Shirley and I later saw, was stored in Tim Jull's office in the Physics Department. Jull was a member of the Arizona AMS laboratory.

Damon said he had consulted a textile expert from Proctor & Gamble about the best procedure for cleaning the cloth—shroud and control samples. There were three controls provided—two from the British Museum (one the British Museum knew the historic age of and the other that had been previously carbon dated) and one from Jacques Evin, head of a French carbon dating laboratory, who had been present at the Turin workshop. Neither the British Museum control samples nor the shroud sample were unravelled into separate threads but the one from Evin was. On

Friday afternoon we saw one of the shroud samples and two of the controls as well as Evin's threads. How Evin had gotten into the act of providing a control, and, for that matter, how he had managed to be present at the Turin workshop, is unclear to me to this day. He was head of a conventional carbon dating laboratory in France but had no AMS or small-counter experience.

Damon described the cleaning procedure employed on one of the four pieces of the shroud sample and about a quarter of the control samples, to be measured the next day. It involved cleaning with a special commercial detergent recommended by Proctor & Gamble. They were then treated with dilute hydrochloric acid to get rid of calcium carbonate and ethyl alcohol to dissolve soluble greases, and finally they were washed in sodium hydroxide. Between each of these steps they were rinsed in doubly distilled water. Linen samples of known age were cleaned in several ways including the P&G way, and excellent agreement was obtained between their known ages and the AMS determined ages. After cleaning the shroud linen, it was still slightly yellowish in colour. The shroud sample could be readily identified. This was actually an advantage because it dispelled the concern that 14th century linen had been somehow substituted for shroud samples. I thought to myself the blind dating nonsense had gone out the door once and for all—but it turned out it had not quite.

Damon described the sample removal in Turin. It had all begun at about 4:30 in the morning on 21 April 1988. The lab representatives had been called to come to the Royal Chapel at about 6 am. Four officials from the Turin city government were at the chapel entrance claiming that permission had not been received from the city to take samples from the shroud. The chapel itself where the shroud was stored, although attached to the cathedral, was the property of the City of Turin and not of the archdiocese. They finally relented when they were invited to witness the procedures. Others present were the cardinal, Gonella, a Turin microanalyst Giovanni Riggi, an Italian textile expert from Turin Polytechnic's Department of Material Science, Professor F Testore (who had asked what this brown patch was when he saw one of the 'blood' stains!), a French textile expert, G Vial from Lyon, Tite and the five laboratory people (Damon, Donahue, Hall, Hedges and Woelfli). The shroud had been taken from its storage place and was unrolled on a table in a small room next to the chapel. There were seating bleachers for the observers on three sides—the fourth side contained the entry door. Riggi was to remove the sample, but it took two hours to decide where it should be taken. Everyone knew it would be near the spot on the hem where Raes' sample had been removed and that is where it was finally cut.

No one handling the shroud, including Riggi, wore gloves. There was absolutely no ceremony—everything was carried out in a businesslike manner. As soon as the sample (a strip approximately 7 cm long by 1 cm wide) had been removed it was weighed. The total weight was about 150 mg. This gives an areal density of 21.4 mg/cm^2 which agrees with the estimate of 22 ± 2 mg/cm^2 mentioned in chapter 6. Riggi then divided the strip into three approximately equal pieces and these were weighed. They were then taken to the adjacent Sala Capitolare where Tite put each sample on a separate aluminum foil which was folded to contain the cloth. He then inserted the shroud samples in their aluminum wrapping in specially machined stainless steel cylinders. One was given to each lab representative along with three other cylinders containing the controls.

Almost the entire procedure had been videotaped and photographed by technicians taking their orders from Riggi. The only part of the operation that had not been viewed by anyone other than the cardinal, Tite, and Gonella was what took place in the Sala Capitolare although most of that was videotaped. This mildly flawed procedure later provided grounds for some 'true believers' to argue that substitutions for shroud cloth had been made. Riggi considered this videotape to be his personal property. The British Broadcasting Corporation had to pay him for a copy and the right to use it in the Cameron film that was presented on their *Timewatch* programme a year later.

Donahue had carried the Arizona samples back to the USA in his briefcase. He went through the metal detector in Turin airport without any trouble. When he took them through customs in the US, he was asked if he had anything to declare. He said "A bottle of gin, some chocolate and a sample from the Turin Shroud". He was passed through by the customs officer with a smile and no questions—much to his relief.

Damon then had to leave to attend some 'Jesse Jackson For President' function. Shirley and I had dinner at the Doubletree and had just gotten back upstairs when the phone rang. It was a reporter named Bill McClellan of the St Louis *Post Dispatch*. Donahue had told me in a phone call I made to him from Rochester before I left that McClellan was married to one of Donahue's daughters. He was visiting the Donahues with his wife and two children and wanted a chance to talk to me. McClellan asked if he could interview me that evening. After consulting Shirley, I agreed we would meet him in the lobby about 9:30. He had done his homework. He had spent a lot of time talking to his father-in-law about the process of carbon dating the shroud. He regularly wrote a column for the *Post Dispatch* about various happenings—court cases, etc—but not science. During the interview he asked me about my quote in some newspaper or other that "Gonella was a second rate scientist" and that he "was a Professor of Metrology—whatever that is—at the Turin Polytechnic". I said that these were correct quotes but that they were somewhat injudicious and impolitic. I would appreciate his not repeating them. He said he would phone me if he had further questions.

The next morning at about 8 am (6 May 1988) I arrived at the Arizona AMS facility. I had asked Donahue to let Shirley attend this historic event since she had been involved in the shroud dating enterprise from the beginning. He said he had been asked to allow the Bishop of Arizona to be present and had turned him down. Under these circumstances he regretted he could not make an exception for Shirley—a deep disappointment for her. I would be the only one present outside the Arizona AMS group. Doug immediately asked me to sign the following statement:

"We the undersigned, understand that radiocarbon age results for the Shroud of Turin obtained from the University of Arizona AMS facility are confidential. We agree not to communicate the results to anyone—spouse, children, friends, press, etc., until that time when results are generally available to the public." It had been signed by D J Donahue, Brad Gore, L J Toolin, P E Damon, Timothy Jull and Art Hatheway, all connected with the Arizona AMS facility, before I signed. My signature was followed by T W Linick and P J Sercel, also from the Arizona facility.

There were three or four members of the AMS team there when I arrived and they had almost finished the five minute per sample cesiation. This consisted of rotating each of the ten samples, located on the ion source wheel, into the cesium beam ensuring that the sample was coated with cesium. During this cesiation process the carbon-12 beam is measured and it was running approximately 12–13 microamps for each. At Rochester we would consider this a very good carbon beam intensity. The samples were rotated into the cesium beam by hand. A stepping motor to do this under computer control was planned but at that time it was done manually and the five minutes were being measured by a kitchen timer. I thought this added a homely touch to the otherwise complex scientific environment. Shortly after I got there they started the second round of cesiation.

I had remarked to Damon the previous evening that I could not think of another scientific measurement that equalled the one about to take place in terms of general public interest—not, of course, in terms of scientific interest. Perhaps the discovery of the tomb of Tutankhamen was in the same class. I made the same remark again and there was general agreement. It was most remarkable for me to have had a major responsibility in bringing the shroud to the test of time, to be about to observe it happen and to learn before anyone but the handful of people present how old the shroud actually was. Damon had been informed that the measurement was about to begin and arrived shortly after. The previous evening he had said he would bet it was 9th century, i.e. 800 to 900 AD or 1188 to 1088 years old. His argument had something to do with when crucifixions ceased.

Eight of the ten samples in this first historic load were OX1, OX2, blank, two shroud and three controls. I did not note what the remaining two were. There may have been some duplicate controls and/or another OX. The OX1 and OX2 are standard samples made from oxalic acid obtained from some crop (perhaps sugar) harvested before and after the atmospheric nuclear weapons tests carried out in the Pacific in the 1950s. The ratio of their carbon-14/carbon-13 ratios was approximately 1.29 and this 29% increase was due to the carbon-14 injected into the biosphere by the exploding nuclear weapons. The absolute values of their carbon-13/carbon-12 ratios was accurately known. The blank was super-pure aluminium and serves as a measure of the background. Arizona's background blanks gave a radiocarbon 'age' in the 40 000 to 50 000 year range. That was such a low background that it could be ignored for samples 2000 years old or less.

Damon said the $1/2$ cm^2 shroud sample being used in this 6 May run had a red silk thread in it as well as some blue threads or fibrils and they had been removed. There was absolutely no problem in identifying the shroud—it was finely, closely hand woven (the weave was not as even as it would have been if done by a machine) and it was the unmistakable shroud herringbone weave.

The first sample run was OX1. Then followed one of the controls. Each run consisted of a 10 second measurement of the carbon-13 current and a 50 second measurement of the carbon-14 counts. This is repeated nine more times and an average carbon-14/carbon-13 ratio calculated. All this was under computer control and the calculations produced by the computer were displayed on a cathode ray screen.

The age of the control sample could have been calculated on a small pocket calculator but was not—everyone was waiting for the next sample—the Shroud of Turin! At 9:50 am 6 May 1988, Arizona time, the first of the ten measurements appeared on the screen. We all waited breathlessly. The ratio was compared with the OX sample and the radiocarbon time scale calibration was applied by Doug Donahue. His face became instantly drawn and pale. At the end of that one minute we knew the age of the Turin Shroud! The next nine numbers confirmed the first. It had taken me eleven years to arrange for a measurement that took only ten minutes to accomplish!

Based on these 10 one minute runs, with the calibration correction applied, the year the flax had been harvested that formed its linen threads was 1350 AD—the shroud was only 640 years old! It was certainly not Christ's burial cloth but dated from the time its historic record began. It was unarguably the least interesting of all possible results. I remember Donahue saying that he did not care what results the other two laboratories got, this was the shroud's age. Although he was clearly disappointed in the result, he was justifiably confident that his AMS laboratory had produced the answer to the shroud's age. Like Donahue, I also had wished for a 2000 year age. That result would have been so much more exciting. Of course, it would not have proved the shroud was Christ's burial cloth but it certainly would have upped the odds. As a scientist, I would have (and did) bet it was not that old but I truly wanted it to be. My exhilaration at being present at this first dating of the Turin Shroud was somewhat dampened by the disappointing result.

When the results of all three labs were finally averaged, the date of the flax harvesting came out to be 1325 AD $\pm$33 years. That agreed with this initial Arizona result obtained in ten minutes using a piece of the shroud cloth measuring less than 1/4" × 1/4" inch. It was a triumph for carbon dating by AMS if not for those who passionately believed it was the burial cloth of Jesus Christ or for those of us who wished it might have been.

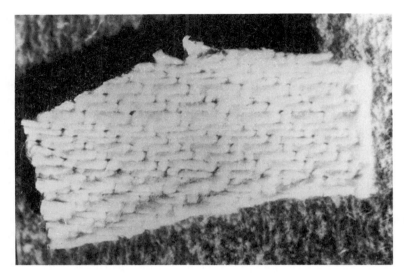

Photograph of one quarter of the shroud sample received by the University of Arizona. The dimensions are about 1/2 cm by 1 cm. The original sample was four times this area. It was divided into four pieces for separate measurements. Note the lack of any contamination. If the shroud were actually first century and modern contamination produced the 14th century result this sample would have to be two thirds shroud and one third contamination.

I had a bet with Shirley on the shroud's age—she bet 2000 ± 100 years old and I bet 1000 ± 100 years. Whoever won bought the other a pair of cowboy boots. Although my guess was wrong, it was closer than Shirley's. She bought me the cowboy boots. The reader, by now, will have guessed that despite the agreement I had signed, I told Shirley the result that had been obtained that day. She and I had been associated with this shroud adventure now for almost exactly eleven years—there was no way I could not tell her. I knew she would never violate my confidence and she never did. Her disappointment in the result was deeper and palpably more poignant than mine. She has told me that, even now, her heart still tells her it is Christ's shroud.

Later in the afternoon of 6 May, Shirley and I had a chance to see the sample in Tim Jull's possession as well as one of the control samples and the separated threads from Jacques Evin's sample. The British Museum control sample was woven with coarser threads than the shroud and Evin's separated threads even coarser than the control. The latter was a fairly loose basket weave. Damon had a micro-photograph of the shroud sample that had just been run, showing the thread from the red silk covering cloth. He had removed all extraneous threads including the red silk thread, so they were not included in the final graphite to be dated.

I had the opportunity that evening to meet Donahue's son-in-law Bill McClellan again at Donahue's house where we had been invited for a drink—Shirley had chosen not to come. It was an interesting evening because, of those present, only Doug and I knew what results had been obtained earlier in the day and what was worse, all the others knew we knew but would not tell. I told McClellan that I would not like to continue my disparagement of Gonella and he agreed he would not quote me in any anti-Gonella way. I said that what I would really like to stress was my very strong view that the pope should have insisted to the cardinal of Turin that he use the Pontifical Academy of Sciences for scientific advice on the shroud instead of depending exclusively on someone like Gonella who was relatively unknown in the science community. I suspected that McClellan would do an excellent job in writing his article for the *Post Dispatch*. He promised to send me a copy and he did.

His article appeared in the Sunday 15 May 1988 edition of the St Louis *Post Dispatch* and was headlined 'Secrets of the Shroud'. The article ended with his description of what occurred after the first measurement was made:

"Donahue was the first person to read the results. The second was Harry Gove, who had been invited to Tucson by Damon and Donahue to observe the test.

"The evening of the day the Shroud of Turin was dated, Gove visited Donahue's home. The two men sat outside on the porch, enjoying the cool spring night. They talked about physics and mutual friends and the desert.

"Everything but the results of the testing of the Shroud of Turin.

"In the coming days the Tucson lab will be testing the other three pieces of its original sample of the shroud. The results will then be sent to the British Museum, which is coordinating the tests. The other two labs will do the same.

"All data will be forwarded to the archbishop of Turin who will announce the results.

"'If the results don't leak out first, it will be a miracle,' said a non-physicist.

"The scientists barely smiled.

"The official announcement is expected later this year."

On the whole it was exceedingly well written and when I read again this last part it still stirs memories of what was an extraordinary experience for me

On 10 May back in Rochester, Ted Litherland called. He said he had received a letter from someone he would not name in Hall's lab in Oxford who told him that the BBC news filmed Teddy Hall taking the sample out of the stainless steel vial. It was used for about 10 seconds on a TV news programme and the picture did not even show Hall—only his hands and this stainless steel vial. The letter from this person in Teddy's lab also said all the 'girls' in the lab were asked to leave early so that it would appear as if it were an all-male operation. I concluded that Ted's letter had come from a woman although I had no way of knowing. I wondered why Hall had felt that was necessary.

On 12 May I flew to London. David Sox had made a reservation for me at the Abbey House Hotel, which was not far from where he lived. It is right opposite Kensington Gardens and the Kensington Palace. I had to climb four flights with my luggage to reach my room but, at least, it was inexpensive. I had dinner with Sox at his place on Friday evening. I now knew the age of the shroud and it was important that I not reveal it to Sox, to Cameron or to anyone else. I did my damnedest not to and I think I was successful.

The next morning, I was to appear at the American School. Cameron had decided that he would film the part of the *Timewatch* programme on the shroud which involved me at David Sox's school, and to make believe that it was my laboratory. I was interviewed by Cameron, as I sat on a stool in one of the chemistry laboratories of this school, about the whole procedure of dating the shroud. There was also a scene in which I was to walk over to the mailbox, pick up a letter and read it. It was supposed to be the one that Sox had sent to me in 1977, shortly after we had done our first carbon dating work on the Rochester tandem. It was all great fun. After that we had lunch at some local pub with Neil Cameron and David Sox and several members of the BBC television crew.

Another evening, I was invited to have dinner at Cameron's house. A woman friend of his was there and it was really a stimulating evening's conversation. Again, they knew that I had attended the actual dating of the shroud in Arizona and that I knew the age of the shroud. They didn't press me too hard, nor did I give them any clues as to what the shroud's age actually was. I left on 16 May and got into Rochester that evening. The total cost of this trip was considerably more than the amount I got from the BBC. However, I had decided it was worth it because I felt that this BBC programme would be one of the really good documentaries on the shroud dating. It would include the film that Riggi had taken at Turin when the sample was removed and it would also show the measurement of the shroud at Woelfli's laboratory in Zurich. I felt that since I had been involved with this thing for so long that I had better be in this programme as well.

On 17 May I drove to Buffalo to appear on the television show hosted by Brian Kahle, on Channel 7 in Buffalo. It was a live morning show with an audience consisting mostly of housewives, I would guess. The programme this morning was devoted to the Turin Shroud. The two other people on the programme were Joe Nickell and Ben Wiech. It was an interesting encounter. Each person was questioned by Kahle. Nickell came through very strongly as a disbeliever that the shroud was Christ's burial cloth and, of course, Wiech as a firm believer. I tried to play a neutral role. Again both Wiech and Nickell were aware that I knew the shroud's age as measured in Arizona.

I later received a letter from Wiech in which he said that I had given no indication whatsoever what the results were that Arizona had obtained.

In a letter dated 23 May Roy Kienitz, an assistant to Daniel Patrick Moynihan, wrote:

"Dear Dr Gove: I am sorry that this had to take so long, but better late than never. Enclosed you will find a copy of the letter that Senator Moynihan sent to our Ambassador to the Vatican. Be assured that I will forward any response that the sender receives to you as soon as it arrives here. Feel free to call here anytime if you have any further needs. P.S. - I heard you talking about this on the National Public Radio last month. Keep it up."

The letter from Moynihan to The Honourable Frank Shakespeare, the US Ambassador to the Holy See, had been sent to him in care of the Department of State in Washington. It read (on Moynihan's stationery):

"Dear Mr Ambassador: A fascinating question for you. It seems that two esteemed research laboratories in New York State, the University of Rochester, and Brookhaven National Laboratory, have been eliminated in the matter of dating the celebrated Shroud of Turin. This decision was taken in the archdiocese of Turin which has been put in charge of the process. As both of these institutions were pivotal in developing the carbon dating equipment and processes the remaining labs will use (left are Oxford, Zurich and

Arizona—yet another case of our innovations being put to the best use by others) it seems only right that we should get some sort of explanation for this disqualification. All of this is detailed in the enclosed letter from Dr Gove in Rochester to His Holiness. Could you bring this up with the right people? Surely not a matter of life or death but of interest all the same. Sincerely, Daniel Patrick Moynihan."

I was very pleased at this response. It contrasted vividly with the total lack of response that I had gotten from the junior senator from New York State, Alphonse D'Amato.

On 24 May I wrote to Sox thanking him for his hospitality and enclosing information about the *13th International Radiocarbon Conference* to be held in Dubrovnik, Yugoslavia, 20–25 June 1988. I said Shirley and I would be attending the meeting and would be travelling there via London. I invited him to have dinner with us while we were in London. I told him that while I was in Heathrow on my way back I had purchased a copy of a recent paperback by Anthony Harris titled *The Sacred Virgin and the Holy Whore*, 'the book that explodes the secrets of religion' according to the cover blurb. Chapter 3 was devoted to the Turin Shroud and in it the author propounded his theory that the original Lirey shroud whose existence was revealed in the year 1355–6 was destroyed and the one we know now was actually painted by Leonardo da Vinci in the period 1490–1500 on linen of unknown vintage obtained from the Holy Land. I said I was sure he would enjoy reading it because, in addition to the shroud chapter, there was a fascinating discussion of Joan, a woman who, Harris claims, later became Pope John VIII in the ninth century.

On 25 May I told Donahue about the Leonardo da Vinci theory expounded in this book. He said that the University of Arizona results were still the same, they had run two more shroud samples. He said that he did not think Zurich would run until just before the meeting in Dubrovnik. He said he really did not care about the date at all, if it were not Christ's shroud. He wondered about the talk that I would be giving in Yugoslavia. I said that the abstract had been submitted some time ago and that I was prepared to give a talk if I were invited. I said I would certainly not mention the age of the shroud in that talk.

I received a letter dated 1 June 1988 from Monsignor Giovanni Tonnucci, Charge d'Affaires at the Apostolic Nunciate to the USA in Washington as follows:

"Dear Professor Gove: Some time ago you very kindly wrote

to Pope John Paul II to express your concerns about the handling of scientific measurements of the Shroud of Turin. The Secretariat of State has asked me to assure you that your letter has reached its destination and your comments have been carefully considered.

"For your information, I am also enclosing a copy of the statement which appeared in the 2 May 1988 issue of the English-language weekly edition of *L'Osservatore Romano*. With every good wish, I remain sincerely yours."

The article was titled 'Samples of Shroud of Turin taken for scientific dating'. It stated that three samples of cloth from the main body of the Shroud were removed on 21 April 1988. The total weight was approximately 150 milligrams comprising a strip measuring about 1 cm by 7 cm. It stressed the procedures followed to ensure blindness and described the three control samples. The ones supplied by the British Museum were stated to be a fabric of the first century AD and the other of the eleventh century AD while a fourth sample, the source of which was not given, was said to be dated about 1300 AD. It gave the names of the two textile experts who were present, Professor Franco A Testore of the Polytechnic of Turin assisted by M. Gabriel Vial of the Historical Museum of Fabrics of Lyon, and said the entire operation was videotaped and documented photographically.

What really surprised me was the fact that the ages of the control samples were given in this news report and they actually corresponded to the results on the three control samples later obtained by the three laboratories. The article appeared even before Arizona carried out their measurements, although I am sure Damon and Donahue were not aware of it (the first Arizona measurement, at which I was present, was carried out six days after the article appeared). However, both Zurich and Oxford made their measurements considerably later and people in those two labs might have been aware of *L'Osservatore Romano* article. I had at least expected the ages of the control samples would not have been revealed to the three labs until after they had submitted their results to the British Museum. In any case, the labs might have been wise to consider the possibility that the ages of the controls quoted in the article were wrong and had been supplied to *L'Osservatore Romano* as a ruse to entrap them.

Shirley and I flew from Toronto to London on British Air on 15 June 1988 and stayed at the Rubens Hotel near Buckingham Palace. We did meet with Sox and on 19 June we flew to Dubrovnik to attend the *13th International Radiocarbon Conference*. The conference headquarters were in a luxurious tourist hotel on the Adriatic Sea.

The meeting went extremely well. My talk on 'Progress in Radiocarbon Dating the Shroud of Turin' was delivered to a capacity audience including Woelfli (with whom I had several long and pleasant conversations during the meeting), Donahue, Damon, Otlet and Hedges. Zurich was very close to having finished their measurements. In his opening talk in the session I chaired, Woelfli was quite critical of some of the experts in the field who suggested that AMS was not really ready to date the shroud and, in any case, there was still not enough experience within AMS laboratories in dating cloth.

Part way along in my talk I stated that, at the invitation of Donahue and Damon, I was present in Arizona when the first measurement was made. I went on to say "Let me tell you the date..." and when I saw the whole audience lean forward with a collective gasp, I smiled and continued "...the date the first measurement was made".

After the talk, Bob Otlet said he was pleased to see that I was not embittered by the experience. He said that he had been approached by Hedges to make the measurement of the stable carbon isotope ratios of the shroud and control samples and wanted to know whether I thought he would be 'letting the side down' by doing so. I said I did not think he would be. Oxford had not yet started to make their measurement because they had had to shift back to their graphite source from the carbon dioxide source they had been working on, so far unsuccessfully, for some time. They were the last of the three labs to complete the measurements and before they did there would be distressing rumours in the British press that the shroud dated to medieval times.

The talk I gave in Dubrovnik was later published in the journal *Radiocarbon* as part of the proceedings of the *13th International Radiocarbon Conference*. I concluded the article as follows:

"The radiocarbon dating of the Turin Shroud which the author had envisaged as a convincing test of the power and efficacy of AMS for carbon dating small samples of precious artifacts, turned out to be a complex and, in some respects, a rather divisive enterprise. It may be that, although there are many questions that science can answer, there are some that it need not and, indeed, probably should not tackle. Be that as it may, whatever age the shroud turns out to be, the result will be contentious in some quarters, in part because of the inadequacies of the procedures being followed. There is a reasonable chance, however, that the three laboratories will independently produce concordant results and, in this circumstance, at least the scientific community is likely to find the dates credible."

Meyer Rubin phoned me on 5 July 1988 to say that he had heard on a Cable News Network (CNN) television programme that a British laboratory had revealed that the shroud was a fraud (i.e. medieval) which he assumed meant 14th to 15th century. I was both astonished and incredulous that either Oxford or the British Museum (these were the only two British laboratories that could possibly be the source of any leak) would have been so injudicious as to perpetrate an untimely disclosure. I called Neil Cameron at his home in London. He said he had heard nothing about this rumour. He said his *Timewatch* programme was due to be aired on the BBC on 27 July and if the rumour seemed credible he would have to rethink the programme. I called CNN in Atlanta, Georgia and was told the news report in question was on the wires yesterday (4 July). They would send me a copy. It said that the British press leaked the story that the shroud dated to the middle ages as measured by Oxford University. It stated that the Vatican believed that no conclusion can be drawn from the results of one laboratory and, in any case, Ballestrero said that the laboratories would not know which of their four samples was the shroud (was he really that naive?). I phoned Doug Donahue but he had not heard the rumour. He thought it was some trick being played by the British press.

The next day, I phoned Hall. His son Martin answered and said his father was out to dinner. He said that Oxford had not yet measured the samples unless they had just done so today. Cameron called me back to say that he had checked with the CNN people in London and they knew nothing about the story. Ben Wiech called me. He had also heard the CNN report on Monday 4 July 1988. He said that he had heard from John Jackson that Peter Rinaldi was very depressed at the news and that the shroud would be moved out of its storage place in the Royal Chapel in a week or so. I thought to myself this was a remarkably hasty and inappropriate response. However, it worried me because one of the people I feared would respond badly to the official news that the shroud dated to the 14th century was Rinaldi. He was well along in years and not in the best of health. How much of a shock would the truth be to him?

The next day I talked to Robert Hedges in Oxford. He said

they had not yet dated the shroud. All they had done so far was to pretreat the cloth. He had no idea how the story got out—obviously the press was unprincipled. He said he was glad Oxford was in the clear. If they had already dated it and found it to be medieval they would certainly be suspect, but that was not so.

Shortly after this call to Hedges, a woman from CNN called. She said that the story had been filed by the Associated Press from Vatican City—it had not originated in CNN. She said she would tell her supervising producer that the story was wrong. I then phoned the Associated Press in New York. I finally got to their foreign editor, Nate Polowitzky, who said that as far as he could tell, the story originated with their AP bureau office in London. They were quoting a Kenneth Rose, writing in the London *Sunday Telegraph*, that rumours had it that the shroud is medieval—Rose did not say where the rumours originated. The AP in the USA did not use the story but it went out on radio and that is where CNN picked it up.

A friend of Meyer Rubin who gets the London *Sunday Telegraph* sent him a copy of the article by Rose, and Meyer sent it on to me. The article stated that scientists selected by Cardinal Ballestrero, Archbishop of Turin, to determine the age of the Turin Shroud had almost completed their task. In spite of the intense secrecy surrounding the investigation there were signals that the linen cloth had been proved to be medieval and not woven in or before the first century AD. Rose went on to note that the number of labs had been reduced by Turin from seven to three which aroused considerable acrimony. He identified the three labs and noted that one of the excluded labs was the Atomic Energy Establishment at Harwell in the UK. He continued that he also heard "unworthy foreign whispers" against Turin's appointment of Dr Michael Tite, of the British Museum's Research Laboratory, as an independent coordinator because it happens that the head of the Oxford Research Laboratory, Professor Edward Hall, is a trustee of the British Museum. Rose commented that the notion that a conspiracy exists between two such distinguished scientists is absurd. He concluded, "Hall, who received a CBE [Companion of the British Empire] in the [Queen's] Birthday Honours, has just taken on an additional job. He has joined Lord Prior and Lord Weinstock on the board of GEC."

Rose certainly was possessed of acute hearing for signals and foreign whispers. He was obviously vastly over-impressed by titles and honours. I privately wondered if the CBE was in any way related to the fact that Hall was dating the shroud. If he were to receive any official award for this activity it would obviously be timely to give it to him before the date was announced, else it might seem to be a reward by the Queen, who is also head of the Church of England, for showing that a relic of the rival Roman Catholic Church was a fake.

On 21 July 1988 Neil Cameron phoned me. He said his *Timewatch* BBC programme was scheduled to air next week and had been so advertised in the *BBC Times*. He did not think the shroud date would be made public until September—the pope would be visiting Turin then. Neil said his peers and superiors thought the programme must give some indication of the age of the shroud. He had concluded, based on information he gleaned while filming in Zurich, that the shroud dated to the 13th century. His supposition was not based on anything that Woelfli did or said. A month or so ago he thought it was 11th century, but since then he was fairly sure he knew the ages of the British Museum control samples so he now concluded the shroud was of the 13th century. He assured me that it was nothing I said during my visit to England in May that led him to his 13th century conclusion. Of course it could not have been since the Arizona measurement placed it squarely in the 14th century. At least Cameron's guesses were moving in the right direction.

He told me that Zurich had just sent their results to the British Museum by express mail. He did not know the status of the Oxford measurements but would find out. He indicated that what he would really like to say on the programme was that the shroud dates to medieval times (i.e. belonging to the middle ages, which is variously defined as about 1000–1400 AD (*Oxford American Dictionary*), 500–1450 AD (*Webster's New 20th Century Dictionary*), and 476–1453 (*American Heritage Dictionary*)). I said that once all three laboratories had sent their results to the British Museum a leak would not be serious in the sense that it would not affect the credibility of the results. It would only be discourteous to the Vatican/Turin. If he leaked information that he appeared to have obtained in Zurich, Woelfli would be very offended. It might also be seen to influence the Oxford result.

Cameron called me back almost immediately to say he had spoken to Teddy Hall in Oxford. Hall said they would start to make the measurements in a week and would send the results to the British Museum a week later, i.e. in early August. Cameron thought that gave him an out—he would say that the measurements were not completed, the results had not yet been obtained by Oxford. That is what he did in the final version of his *Timewatch* programme.

Sox sent me a copy of an article he had written for the *BBC Times* advertising Cameron's *Timewatch* programme on carbon dating the shroud to be shown on BBC 2 on 27 July 1988.

"What will the verdict be?   Most observers anticipate a medieval date or one near the time of Christ. Few of the shroud enthusiasts or the faithful are prepared for the former. After all the shroud has had a good run. ...After reading all the material in the shroud's favour a London vicar said to me 'God doesn't operate in this manner does he?' I don't think he does but one thing is for certain unlike all other examinations of the shroud carbon dating gives a date."

Sox also sent me the article he had written for the 28 July 1988 issue of the Catholic publication in Britain, *The Tablet*. It was titled 'Focus on the Shroud' and led off with the following summary:

"The results of the crucial carbon dating of the Shroud of Turin are about to be made known. Will they confirm or refute the claim that this could be the burial cloth of Jesus? An Anglican priest who was formerly the general secretary of the British Society for the Turin Shroud looks at some of the implications."

He recounted the trip he had made to Turin on 20 June 1988 during which he attended mass in the Royal Chapel where the shroud rested in its reliquary. He described his old friend Peter Rinaldi, 'Mr Shroud', and how he gingerly asked him "What if the Shroud turns out to be a fake?" Rinaldi's answer surprised him. "A date is a date. We should prepare for that possibility." (Maybe the suggestion Sox had once made to me that Rinaldi had somehow known for some time that the shroud was medieval was true after all. But, if it were, how could he have known?)

Sox described his involvement in Cameron's *Timewatch* programme for the BBC and his visit to Zurich with Cameron and the BBC film crew and how impressed he had been with Woelfli and the ETH tandem operation there. He particularly stressed how impartial and unbiased the senior scientists at the three laboratories were and how "any talk in the future of any kind of collusion will be plain silly". He wrote, "If the date is given as near the first century, there will be no stopping the enthusiasm". In that event "Turin will be 'over the moon' with excitement. I even heard one priest suggesting that his city would become 'a second Mecca'." (As mentioned previously in this book, I heard the suggestion that a monument rivalling Napoleon's tomb should be built to house the shroud.)

Sox concluded his article "I am not sure I want the Shroud to receive a first century date, for fear of what unbridled enthusiasm might do to it. ... It is an icon—and a very good one."

Ian Wilson and his wife visited Rochester on 5 August and I took them to dinner at the University of Rochester's Faculty Club. Wilson was mostly interested in probing the origins of the leaks concerning the medieval age of the shroud. I am sure he believed they came from information I had supplied to Sox. He brought me a videotape copy of Cameron's *Timewatch* programme 'Shreds of Evidence'. It was on the British TV format which has more lines to the inch than US TV. Shirley and I later watched it on a machine designed for the British format in the university's visual equipment centre. I subsequently got a copy from Doug Donahue that he had had converted to the US format.

It contained scenes of the actual sample taking videotaped by Riggi that the BBC purchased from him. At one point a long table covered with a royal blue cloth was shown. As the cloth was slowly folded back the video camera revealed, lying on the table, an incredible display of gleaming instruments including many different kinds of scissors, tweezers and other unidentifiable tools worthy of a brain surgery operating room. A little later Riggi was shown cutting the shroud. The casualness of this operation was emphasized by the fact that he was not wearing gloves! So much for the sterile environment and procedures in handling the shroud Riggi had emphasized at the Turin workshop. It also covered in some detail the actual measurement being made on a shroud sample on the tandem in the ETH lab in Zurich— Willy Woelfli presiding. It also showed most of the scenes of me that were filmed in London in May 1988. It ended with the statement that Arizona and Zurich had submitted their results to the British Museum but that Oxford had not yet completed their measurements.

Meanwhile, the story that the Shroud of Turin was a fake was getting increased attention from the press. The original rumour that the shroud was medieval appeared in the article by Kenneth Rose in the London *Sunday Telegraph*. Aside from a naive statement from Ballestrero that the labs would not know which of four samples was the shroud, there was not much reaction to the Rose report. However, this changed when the 27th August 1988 edition of the *Washington Post* carried a story by Tim Radford of the *Guardian* that "The furor began after Dr Richard Luckett of Cambridge University wrote in the *Evening Standard* yesterday that a date of 1350 'looks likely' for the 14-foot piece of linen which appears to bear the imprint...of Jesus. He also referred to laboratories as 'leaky institutions'." The *Washington Post* reported they had been unable to contact Luckett for comment nor could they get any response from Hall's lab in Oxford or from Tite at the British Museum except for confirmation that the results from all three labs were now in.

Somehow the impression had been created that the 'leaky institution' Luckett referred to was Hall's Oxford Laboratory because the *Washington Post* quoted Gonella as saying "Frankly we in Italy feel we have been taken for a ride. I am amazed that there should be indiscretions of this sort from a university like Oxford. We had expected different behaviour from a laboratory of this reputation." Ballestrero was also quoted: "[It was] all nonsense. It is a blind test and no one in Britain has the key to identify the samples." The *Post* revealed the naivety of this comment by pointing out that the shroud had a very distinctive—and rare—weave and that a scientist who knew anything about the shroud would know which sample he was handling.

A friend of mine who was visiting Mexico sent me a clipping from the 27th August edition of the *Mexico City News*. It quoted the report carried by the *Evening Standard* on 26 August and provided a few more details from that report. The *Evening Standard* report claimed that Oxford had found the shroud to be a fake which dated only to 1350 AD. It gave no attribution for its report but quoted Dr Richard Luckett of Magdalen College, Cambridge as saying "I think that as far as seems possible the scientific argument is now settled and the shroud is a fake". This time it was Gonella who was quoted as saying the tests were blind and that "[Oxford was] asked to date the three [sic] samples, not to determine which of them belongs to the shroud".

Someone by the name of David Boyce, writing in a monthly publication of the *Catholic Counter-Reformation in the XXth Century*, directly accused Hall of this leak. The article was referred to in the *Evening Standard* of 26 August 1988. Oxford had completed their measurements during the first week of August and had sent them to the British Museum. Hall certainly knew the Oxford result at the time of the leak and may also have known the overall result that was to be published in *Nature*. Both gave a mean several decades less than 1350 AD. Hall had no motive for perpetrating the leak and the clear disparity between what he knew the answer to be and the leaked date is convincing evidence that he did not.

The Rochester *Democrat and Chronicle* also carried the story on the front page of their 27th August edition under the headline 'UR (University of Rochester) scientist rejects story of relic's age'. The subhead read 'London paper claims tests show Shroud of Turin a fake'. The report read: "The University of Rochester physicist who is a major pioneer in the newest method of dating ancient materials last night rejected a British newspaper report that the Shroud of Turin has been found to be a fake. ...Last night Dr Harry Gove...said that no information has been released on the tests. European experts also discounted the newspaper report as an unsubstantiated rumour. The London *Evening Standard* yesterday reported, without attribution, that radio-carbon tests at Oxford University showed the shroud was made about 1350. That report, Gove said, is 'completely incorrect'." I was quoted as predicting the final results would be available perhaps as soon as early September. Susan Black of the British Society for the Turin Shroud was quoted as saying "This report is without any serious foundation".

The article went on "At Oxford's Research Laboratory for Archaeology and the History of Art where the tests were made, a woman on the staff who would not give her name said 'No data has been disclosed by this laboratory. I presume the director knows the results but the rest of the staff don't even know and we are sworn to secrecy anyway. The only people who know are the director and professor of the laboratory and they are hardly likely to disclose it'."

The article stated that Luckett, whose university is an ancient rival of Oxford, was not connected with the tests but had been associated with investigations of the shroud's history. "He wrote in a separate article in the *Evening Standard* that laboratories 'are rather leaky places' but did not elaborate." It then quoted Gonella as having been amazed at the suggestion that Oxford had leaked this information.

The same edition of the *Daily Mail* carried a rather poignant and touching article written by Group-Captain Leonard Cheshire VC (the Victoria Cross—the highest military award for valour bestowed on members of the armed forces of countries belonging to the British Commonwealth). Cheshire had long been associated with the British Society for the Turin Shroud and was its vice-president when he wrote this article. It was headlined 'Why I still believe in the Turin Shroud'. He referred to the leaked report that the shroud dated to the 14th century and wrote "There are things that science can do. There are places where science cannot venture. The eternal mystery of the resurrection is one of them." Cheshire was founder of the Cheshire Foundation homes and co-founder of the Ryder Cheshire home for the disabled and infirm at Cavendish in Suffolk where he lived. He wrote that for him after the 35 years he had known the shroud "it speaks in the language of today. It tells the same Gospel story, of Passion and Death and Resurrection, but through an image clear to us all. It makes the Passion more compelling, more personal. It tells of a face that has suffered and loved. Science cannot alter that." It reaffirmed my feeling that, although there are questions that science can answer, there are some it should not even try to answer. Group-Captain Leonard Cheshire VC would say that there were some questions that not only should science not try to answer but ones it inherently cannot answer. He believed the mystery of the Turin Shroud was one of them. He is a man whose courage and good works speak for themselves. I am glad the Shroud of Turin still strengthens his faith despite what science claims to have revealed about its age.

An Associated Press story appeared in the 9 September 1988 issue of the Rochester *Democrat and Chronicle* headlined 'Shroud's age remains secret Oxford research chief says', with the subhead 'He claims forgery report was just a guess'. Teddy Hall was quoted to this effect in the *Oxford Mail*. The article went on "But Dr Richard Luckett, a Cambridge University professor, said he stood by his word, adding, 'I had an absolutely marvellous leak from one of the laboratories and it wasn't Oxford.' Luckett, last month, said tests at Oxford showed the shroud was made in 1350. Hall condemned the statement as 'mere guesswork,' saying 'The person who has been doing the dating told me that Dr Luckett got the date completely wrong'."

I must say I wondered about Luckett's date of 1350 because it was the date Donahue announced to me when I was present at the first radiocarbon measurement on the shroud in 6 May 1988. Of course, it also corresponds very closely to the shroud's known historic date. However, I still assumed Luckett had said he got the number from Oxford. When I read that he claimed he got it from one of the other two labs I worried that it might have come from someone who was present at Arizona during the first measurement. However, it did not really matter now since all three labs had submitted their results to the British Museum and so none of them could be influenced by this real or imagined leak. Shirley convinced me that it was, in fact, a guess as Hall had stated. After all, the historic date for the shroud was circa 1353 when de Charny founded the church in Lirey, France purportedly to house the shroud.

I had another worry, however. The *Oxford Mail's* report of Hall's response (that Dr Luckett "got the date completely wrong") to the 1350 date seemed to indicate that Oxford might have come up with a number quite different from Arizona. I phoned Donahue that same day (9 September) and read him the *Democrat and Chronicle* report. Donahue said all three labs were in agreement and the data had been sent to the Institute of Metrology in Turin. He said that Tite thought an announcement could be made before the middle of October. I was considerably relieved. My worst fear was that one of the three labs would again foul up the measurement and thus bring accelerator mass spectrometry into disrepute. Mercifully, it was not to be realized. I have to confess to feeling that if any of the three labs did make a blunder in their measurement, it would have been divine justice if Oxford had been the culprit.

The mystery of how the rumour of a 1350 date for the shroud had started deepened. Articles that appeared in the 21 September issue of the Rochester *Democrat and Chronicle* from the United Press and the 22 September issue of the *New York Times* by Malcolm W Brown both referred to reports in the *Times* of London variously stating that the tests revealed a date between 1100 and 1500 and about 1350.

They quoted Dr Robert Dinegar, an Episcopal priest who is also a professor of chemistry at the University of New Mexico at Los Alamos, as saying he knew what the results of the carbon dating measurements were—he claimed he had been told the date of the shroud by an authoritative source. (Who was that source I wondered, or was Dinegar bluffing?)

Amusingly enough he made it appear that he was somehow intimately involved in the carbon dating effort (which, of course, he was not in any way) but could not divulge the results because all of those involved in the project were bound by signed undertakings not to reveal the results of the analysis until they were announced by the archbishop of Turin or his scientific advisor, Luigi Gonella. He broadly hinted in an interview with the *Los Angeles Times*, however, that the leaked result of 1350 was not far off the mark. Dinegar continued, quite improperly and completely at variance with the facts, to publicly assert his involvement with the carbon dating of the shroud. One would expect better from a man of the cloth that he apparently had become.

The article in the *New York Times* ended: "According to the *Times* of London report, the dates determined by all three laboratories were close enough together to confirm each other and to establish the date of the shroud as about 1350".

David Sox called me from London on 23 September 1988 to say that Ian Wilson had charged him with being the source of all the leaks. Sox vigorously denied the charge. On 27 September I phoned Donahue. He told me that Damon had phoned Gonella and that Gonella had conceded that the rumours were correct but that it was not yet official. Gonella also told Damon that he believed the rumours came from me to Sox. Damon himself believed that Sox was the source of the leaks. I assured Donahue that I did not tell Sox the date. He probably knew as much as Cameron did as a result of having been in Zurich during the BBC taping of their shroud run. Doug said the results had been sent by courier from the British Museum to Cardinal Ballestrero in Turin and he probably had them now. It was up to the cardinal. He said all three labs were in agreement on the dates for the shroud and the three control samples.

David Sox phoned me the same day to tell me more about what he referred to as 'a pastoral visit' to Peter Rinaldi he had made on 20 June. By this time, Sox was so convinced that the shroud's date was medieval that he felt he should break the news to Rinaldi in person. I suppose he felt, as I did, that it would be a tremendous shock to the man who, more than any other, had been associated with the shroud all his life. Rinaldi told him that he had been prepared for this for the last six or seven years because around that time a secret carbon dating had been made of the shroud!

Meanwhile, reactions to the rumour that the shroud date was medieval continued. On 1 October 1988 the Toronto *Globe and Mail* carried a story by John Allemang headlined 'Losing Faith— Catholics are rocked by revelations that the Shroud of Turin is a fake'. The 3 October 1988 issue of *Newsweek* carried a story 'Defrocking the Shroud—Only a medieval hoax?' It stated that the results of the carbon dating had been complete and "the results are leaking profusely". London's *Evening Standard* had reported in August a date of 1350, which had been endorsed by Cambridge

University historian Richard Luckett. He claimed the investigators had shared their results with him independently.

On 18 September London's *Sunday Times*, citing an unnamed source, reported that all three labs had independently placed the age of the linen in the same period of medieval history. It concluded that "The shroud is undoubtedly the work of a brilliant medieval hoaxer". Gonella described the reports as 'common gossip' and 'propaganda against the Catholic Church'. The Arizona team refused to comment and Hall of Oxford described the reports as guesswork. However, University of New Mexico chemist and Episcopal priest Robert Dinegar, who, the story said, had helped organize the study and had privileged access to all the labs' findings, hinted broadly that the stories had been on target. "I've been involved in this investigation for many years. The scoops always come from England and they almost always turn out to be correct."

I was absolutely fascinated by these marvellous leak stories. It made me feel that there was some possibility that someone connected with one of the three labs or with the British Museum was providing confidential information to Luckett, Dinegar and various newspapers. There was no harm done at this stage since the lab results were in the hands of Tite in the British Museum before the leaks began, so the data were not compromised. I could imagine the possibility that someone at Oxford or the British Museum could have leaked to Luckett—a man I had never heard of before or since—and to certain British newspapers. What their motive might be was unclear. But how Dinegar, whose STURP connections made him *persona non grata* to all the senior people in the three laboratories and the British Museum, could possibly be privy to any confidential information concerning the carbon dating results was impossible, unless, of course, Gonella had told him. Perhaps it was all a result of lucky guesses. We will probably never know. However, there was more to come.

On 7 October 1988, Damon phoned me to say that the official announcement would probably be the next week. He said that Gonella had been told by Rinaldi that the rumoured date came from me to Sox and then to Rinaldi and the world. He claimed that Hall was ready to announce the results if Turin delayed the announcement much longer.

I then decided to phone Gonella. I told him in no uncertain terms that he had better be damn careful about ascribing the leaks to me. So far as I know he stopped doing so. That was my last conversation with Gonella.

On 12 October I received by express mail a copy of a letter dated 7 October and postmarked 8 October 1988 that Sox had written to Rinaldi.

He said he was both angered and saddened by Gonella's suggestion to Paul Damon that he (Sox) had told Rinaldi the date of the shroud test and that Harry Gove had told him. That was not true. He did not know the date. He had no knowledge of the Cambridge professor who 'leaked' to the *Evening Standard* and no one he knew did either. Sox went on to say in this letter to Rinaldi that Gonella dislikes Gove as is well known, but "to colour the release of the results with acrimony is both silly and beside the point. Given this extended STURP-like delay of releasing information leads to this type of acrimony. And this ganging up on long-time skeptics will make Turin look very foolish in the long run. I told you my doubts as a priest to a priest thinking (misguidedly, I now suppose) it was better to hear from me than anyone else. Woelfli did not tell me, Gove didn't and I will never say how I came to have an inkling about the results but I have had them a very long time especially after things King Umberto intimated to me—and he could not bring himself to say to you."

This letter settled the question of whether I revealed any information to Sox regarding the results obtained on the shroud's age during the first run I attended at Arizona. When I was in Britain in May to participate in the BBC's shroud programme I had taken great care not to tell him or Cameron or anyone else the results I had observed first hand in Tucson. But it also raised a number of questions. It implied that King Umberto knew more about the origin of the shroud than had ever been made public and suggested to me that it would be well worth someone's time in going through his private papers if that were allowed—something I had suggested before. It was not out of the realm of possibility that the name of the artist would be found in these papers if they were ever made public.

Some time in mid-October 1988, I received a document dated 7 October 1988 written by Meacham. It presented his concerns about the possible exchange of carbon isotopes that could have taken place during the fire in 1532. This and other forms of contamination of the small samples that were taken from a single location on the shroud had been constantly pointed to by critics of carbon dating as reasons not to trust any shroud date so obtained. Remarkably enough, the statement was made by Meacham that the sample was taken from a scorched area. (That was simply not true as Meacham very well knew.) The document also quoted Dr Stuart Fleming of the University of Pennsylvania as saying that a skilled medieval restorer "could have re-woven a damaged edge (where the sample was taken) to a standard not visible to the naked eye". (The shroud samples used for the carbon dating, of course, were subjected to examination by the laboratories involving somewhat more sophisticated devices than the naked eye and were clearly not the result of 'invisible mending'.) What astonished me most, however, was the inclusion in this document of a copy of the letter I had written to Sir David Wilson, director of the British Museum, on 27 January 1988. It was offered as evidence of my belief that the tests would be worthless. How Meacham obtained it is a mystery to me and his misuse of it confirmed that he, like Gonella, was willing to toss principles aside when it was convenient. All in all the 'Meacham' document was scarcely likely to be taken seriously by anyone with more than a rudimentary knowledge of science and its workings.

The 8 October 1988 issue of *Science News* carried a story 'Shroud of Turin is fake, official confirms'. It reported that Luigi Gonella, scientific advisor to the archbishop of Turin, had confirmed the rumours that the shroud dated to the 14th century. It stated that this was a vindication of conclusions reached by Walter McCrone.

Ben Wiech, the Buffalo lawyer, with whom I communicated from time to time, sent me a news release dated 10 October 1988 issued by William Meacham. It quoted him as saying "The sampling procedure was seriously flawed, and this round of testing has proven nothing about the shroud as a whole. The samples were taken from a repaired corner that had also been scorched in a fire in 1532."

Fascinating new information concerning the alleged carbon dating of the shroud six or seven years ago was also provided in another release by Meacham, dated 14 October 1988. It stated that "A closely guarded secret testing of the Shroud of Turin in 1982 by an American C-14 laboratory yielded conflicting dates of 200 AD and 1000 AD, an American archaeologist involved in Shroud research claimed today. William Meacham of the University of Hong Kong said that members of the US scientific team (i.e. STURP) informed him that a single thread was...tested at the University of California nuclear accelerator facility. He said separate ends of the thread gave quite different results... These results were never published because C-14 testing did not have the approval of the Turin authorities at that time."

Meacham claimed that Gonella was only informed of the results in 1986 at a meeting that Meacham had attended. Meacham also repeated his claim that the sample dated by the three labs came from a corner of the shroud that had been scorched in the fire of 1532 and that it was possible that this area of cloth was re-woven by a medieval restorer. He concluded this news release by citing a letter he had just obtained written by one of the lab directors (unidentified) as describing the current C-14 project as "a rather shoddy enterprise". This, of course, was another reference to the letter I had written to the director of the British Museum on 27 January 1988 and referred to above and in the previous chapter. He had no right to have a copy of the letter or to refer to it out of context.

I had dealt with the assertion that the samples from the shroud were scorched and/or re-woven but how about the possibility that a scientist at a California nuclear accelerator facility had secretly dated each end of a single thread from the shroud in 1982? The only accelerator anywhere in California that had ever been used for AMS was a cyclotron at Berkeley. The person involved in that attempt was a scientist named Richard A Muller. There was not the remotest possibility that Muller could have or did date the shroud in 1982 or any other time and the absolute proof of this was the fact that he never claimed to have done so. Meacham either was being duped by someone or was fabricating the facts. In my opinion both Meacham and Dinegar had been out of their league concerning the Turin Shroud since 1978. Their attempts to stay in the running would be sad if they were not so annoying.

Chapter 11

Results Announced, Reactions

On 13 October 1988 the official announcement came from Cardinal Ballestrero, Archbishop of Turin. The flax from which the shroud's linen was woven had been harvested in 1325 ± 33 AD. This was very close to—and within the small uncertainties of the measurement—the historic date of circa 1353. There was complete agreement among the three labs, not only on the date of the shroud samples but on the three control samples as well. It was a triumph for AMS if not for the millions who believed that this extraordinarily beautiful cloth with its poignant image of a crucified man was the burial cloth of Jesus Christ. Although I had known for five months that some date close to this was the answer, I still felt pangs of regret. Regret that I had been largely responsible for this iconoclastic result and regret that there was now no possibility that the Turin Shroud was Christ's burial cloth. I had really hoped in my heart for a first-century date although my mind told me that was bordering on the impossible.

The *Times Union* of Rochester, an evening paper, in its 13th October edition carried a front-page story taken partly from the Associated Press and partly from interviews by Marie McCullough, a *Times Union* reporter. The story was headed 'It's official: Shroud not over 728 years old'. (The publication would state there was a 95% probability the flax had been harvested between 1260 and 1390 AD. Adding 728 to 1260 gave the publication year, 1988. The 95% refers to what is called a 2 sigma uncertainty in statistics. It is more usual to quote a 1 sigma uncertainty and that means there is a 66% probability that the harvest took place between 1292 and 1358 AD or 1325 ± 33 AD.) It quoted Ballestrero as saying he saw no reason for the church to doubt these results. It said Luigi Gonella bristled at the suggestions that the shroud was a forgery. He pointed out that the word 'forgery' implies that someone created it deceitfully and there was no proof of that—it could be a medieval icon. There was still a mystery as to how the image was formed.

The report stated "University of Rochester Professor Harry Gove, who developed the carbon dating technique used on the shroud but was excluded from the testing process, said today that the unanimous result was a 'triumph' for the technique. 'The fact that these scientists could take such an exquisitely small sample and get an age for it—and the fact that the technique has matured to the point where all three laboratories get the same answer—that says to me this [technique] has come of age.' He added wryly that the 14th century date is 'probably the dullest age they could have come up with. If it were a bit younger maybe it could have been painted by Leonardo da Vinci,' Gove said with a chuckle. Gove said he hopes the scientific finding will not shake the faith of those who believe the cloth is a sacred relic. 'There are people I've met during the whole marvellous adventure whose faith was substantially changed because of the shroud. I know some who changed from being agnostics to Christians,' he said. 'I hope their newly found faith won't be shattered by this. I know it will be an enormous disappointment to the people in Turin for whom this was an important relic and something that brought prestige to the city.' Now, Gove said, Catholic Church officials should invite art experts to examine the cloth to explain the many unanswered questions about how it was created."

On 14th October the *New York Times* covered the story on page 1 with the headline 'Church Says Shroud of Turin Isn't Authentic'. It quoted Cardinal Ballestrero as saying "These tests do not close the book on the Shroud. This is but another chapter in the Shroud's story, or as some would say, in the mystery of the Shroud." Ballestrero pointed out that "after all this research, we do not have any plausible answers to explain how the image was created". The article emphasized that far more extensive tests than any carried out so far were needed to solve the puzzle of the image. Ballestrero claimed that the shroud had, however, produced miracles and would continue to do so. (This was the first time I had heard that claim made—in fact, to the contrary, I had been told that no miraculous cures had ever come to people viewing or even touching the shroud.)

Ballestrero said that when he informed Pope John Paul II of the result, the pope had said "publish it". Gonella said he was satisfied that the results from the three labs coincided to produce a date of 1325 AD for the time the flax from which the shroud was made was harvested and that proper procedures had been followed to rid the cloth of material that could have distorted the dating. Both he and the cardinal expressed their annoyance that unknown members of the research effort had violated a voluntary secrecy agreement by disclosing the test results to the press in August.

Whether the leaks were a result of educated guesses, which I now consider most likely, or actually came from someone connected with the three laboratories or the British Museum, did not really matter in the end. There was no way in which they could have biased the final result. All the data were in the hands of the British Museum when the leak floodgate really opened with the Cambridge University historian Richard Luckett's revelations that a leak to him from one of the laboratories gave a date of 1350 AD. This occurred on 27 August 1988 and Oxford, the last lab to complete its measurements, had submitted its data to the British Museum in early August. It was true that a reporter for the London *Sunday Telegraph*, Kenneth Rose, in an article dated in early July before Oxford began its measurements wrote that he had "heard signals" that the date was medieval. Anyone claiming that this tenuous hearsay could have influenced the Oxford measurements would have to be psychotic.

Tom Beasley, a co-worker in the field of AMS, sent me a clipping from the Chicago *Tribune* of 14 October written by Kenneth R Clark who, in recent years, had often written on shroud matters. In a note accompanying the clipping Tom asked "Now that it's done how do you feel?"

Clark's article was headed 'Carbon tests prove shroud is not burial cloth of Jesus'. After reporting the date that had been obtained by the three labs and Ballestrero's reaction to it, the story continued "The Vatican statement, which had been expected for weeks, raises more questions than answers, posing a classic 'whodunit' surrounding the shroud, which records show first appeared in France in the 1350s and had been in Turin since 1578. If the shroud actually is the work of some unknown Renaissance artist, he has had no peer before or since."

I thought that statement was a bit of an exaggeration. Had Clark forgotten Michelangelo, Leonardo da Vinci, Rembrandt and so many others? For that matter the artist might have been a woman.

The article went on to describe what the artist would have had to accomplish—blood but no paint, it was a photographic negative and it had three-dimensional properties. It then gave the comments of James Druzik, of the Getty Conservation Institute in Marina del Ray, California. This was the same Druzik who had been listed as one STURP member expected to attend the Turin workshop in 1986, but never did—and a person about whom I had some reservations. If STURP was proposing him he must not be a good choice and besides Getty was not high on my list of admirable people. It turned out that I was wrong about Druzik.

Druzik was quoted as saying that the image might be less a work of art than an accident. He said "It still befuddles one on how it could have been made, but in this particular field, there is a lot of befuddlement on how things were made. There are many, many articles that have transfer stains on their surfaces that were unintentional. Conservation literature is full of stains that occur from decomposition in products from wood to paper or textile. It's so common that it's not even remarkable." The article stated that Druzik was a later associate of the American scientific team that ran the Turin tests (i.e. STURP). He said the image could be one of, not a corpse, but a statue—a carved wooden icon, of which there were many in the 14th century. He noted that people at that time had virtually no education by our standards. Even if one were a priest, one's education was probably limited to theology, so "you were spooked by anything you didn't understand. If you had a piece of linen draped over a wooden icon, and you came back 10 years later and found an image appearing on it, you'd be totally freaked out. With the first word-of-mouth transmission, it goes back 1000 years." The article stated that Druzik even has his doubts as to whether all the scientific tests run in 1978 and earlier were relevant. He said he took the negative-image and the three-dimensional aspects 'with a grain of salt' as possibly coincidental, but he did concede that no paint pigment of any kind was found on the shroud, making it difficult to believe it was a painting. He said of the scientific team "They had very little to work with. They strived with amazing creativity to find every conceivable approach to deriving information as to the origin and structure of the image. But the problem of looking at large objects with small-object technology is that it's like looking through a keyhole and trying to determine the character of the person who lives in the room."

It seemed that I had clearly misjudged Druzik. It was hard

to believe he really had STURP connections and could still talk so sensibly and dispassionately. There was nothing in any of the above statements with which I could quarrel except possibly his description of the STURP research as creative and his ignoring the claims of Walter McCrone that applications of iron oxide could have produced the image. As an art conservator, he would make an excellent member of a team of art experts that should be assembled to examine the shroud—only if he were willing to shed his STURP connections, of course.

The article stated that whatever the final conclusion, Dr Harry Gove of New York's University of Rochester, a physicist who initially was critical of the way carbon-14 tests were administered, said the shroud was almost as precious now as it would have been had the tests authenticated it as a true artifact of Jesus. I was quoted as saying "If it isn't a relic—Christ's shroud—it could be an icon, an artistic representation that would make it the most important piece of Christian art the world has ever known, that's no mean fate. They should let the art experts look at it now. To pin down who actually did it and how it was done would be a remarkable thing."

The article continued, "Gove, whose lab specializes in the [AMS] process was to have been one of those making the tests. Since carbon-14 requires destruction of a small portion of the object being tested, however, the church backed away from an original seven-lab agreement, cutting the number to three and raising doubts as to the statistical accuracy of such a test. Gove was one of the harshest critics at the time, and he conceded this week that while most scientists—himself included—will accept the results, there were enough faulty procedures to raise doubts. 'The piece that was removed from the shroud was divided among the three labs and that piece came from one specific spot in the shroud,' he said. 'If there were some reason why the carbon-14 content in that particular piece was contaminated, it's inaccurate. All of the labs used the same cleaning technique, and if there's some kind of contaminant that was not taken care of, it would give the same answer to all three labs, and all three would be wrong. There's an out for those who want one, but I think it's so unlikely as to be impossible. I'm also very confident there has been no intercommunication or collusion between the labs. They've been enormously circumspect in that area'."

I gave more thought later to the question of contamination affecting the date. It is easy to calculate how much was required, depending on the time the contamination occurred, to change the apparent date from first century to fourteenth century. I concluded there would have to have been so much contamination it would have been obvious to the naked eye.

In a letter dated 9 November 1988 I responded to Tom Beasley's question of how I felt now that it was all over:

"Dear Tom: Thank you for sending me that recent clipping from the Chicago *Tribune* and all the previous ones you have sent. I particularly enjoyed the comments of Druzik of the Getty Conservation Institute. I had my suspicions about him because he joined the STURP team a while back. His remarks, however, made eminent sense to me. Art experts like him should now be given the opportunity of examining the shroud to find out how the image might have been created. It is also amusing to read that Luigi Gonella, the Cardinal of Turin's science advisor. . .now thinks the shroud might be an icon. Several months ago I suggested in a letter to him that, if the shroud was substantially younger than 2000 years, it might be the most important piece of Christian art and that would be no mean fate for the shroud. . .

"As for me, although if I had had my druthers [my preferences] I would have liked it to be 2000 years old, I consider the results a triumph for tandem AMS. It is certainly a dramatic example to the public of the efficacy of this new carbon dating technique. When I was in Arizona in May when the first measurements on an exquisitely small shroud sample showed in about one minute that it dated to about 1350 AD, even as a co-inventor of the method, I was mightily impressed. Beyond that, however, I am relieved it is over. It was a marvellous adventure while it lasted, despite some obvious disappointments, but it really lasted too long. Yours sincerely, Harry."

In a note dated 24 October 1988 from the irrepressible Meyer Rubin, he wrote "Harry—I don't think you get the morning Naples paper, so I'm sending you my copy. I was in Naples checking out [Mount] Vesuvius when the news came out. I tried to see the Pope in Rome. Said I was a friend of yours. Those Swiss guards can be rough! Best, Meyer, the Pasta Kid." There was a footnote "you do read Italian, of course".

Peter Rinaldi sent me a letter postmarked 21 October 1988 from Turin. In it was a postcard of the hologram of 'The Holy Face from Holy Shroud and The Face of Christ by Aggemian (1935)'. Meyer Rubin had sent me a stack of calling card versions of this same hologram some years ago and I had given them to various people including Professor Chagas. On the back of this postcard Peter had written "Dear Harry: I pray this brief note finds you in good health and spirits. See enclosed messages for an update and recent Shroud developments. More about this over the phone when I get back (before Christmas sometime) in Port Chester. Kindest regards! Peter."

Enclosed were two messages from Rinaldi on the letterhead of the Holy Shroud Guild of which Otterbein was president and Rinaldi vice-president. The first was dated 15 October 1988, just after Cardinal Ballestrero made the official announcement that the shroud dated to the year 1325. Peter wrote to his shroud friends that this had been a busy and difficult time for him. "I doubt if the cause of the Shroud ever went through more trying times than it did during the last year. Surely you must know by now that the scientists, using the ultimate test, the carbon-14 analysis, have dated the origin of the Shroud to the 14th century AD. This would mean, of course, that the Shroud is not the burial cloth of Christ."

Did Peter really believe that? Had he known that for some time? That is something I would really like to know—and something I will probably never know.

He continued: "Let me say, first of all, that not all the experts accept the results of the test. Some of them are actually calling for a new test on good scientific grounds. I was intrigued by what one of them told me: 'Valid or not, the results of the carbon-14 test in no way solve the mystery of Christ's image on that cloth. The test has not said the last word on the Shroud'."

He recounted how, "Shortly after the results of the carbon-14 tests were announced, a friend met me in front of the Turin Cathedral. Placing his hand on my shoulder, he said mournfully: 'I feel terribly sorry for the Church and for you'. 'You can't be serious,' I told him. 'Do you really think the Church will fall apart because the Shroud may not be what many of us supposed it to be? The Church has nothing to fear from the truth, provided, of course, it is backed by solid facts. ...I might be persuaded to accept the results of the test only when someone will demonstrate beyond all question, how a medieval artist produced so extraordinary an image as that of the Shroud'."

Rinaldi wrote that for him the image on the shroud would continue to be, in the words of Pope John Paul II, "a unique and mysterious object, its image a silent witness to the passion, death and resurrection of Christ". As for him Rinaldi wrote, "as often as I glance at the image of the Man of the Shroud, my heart still says: 'It is the Lord'." He was, however, worried that the simple faith of many good people may be shaken by this turn of events. He counselled them to remember that it is the Lord that matters, not the shroud.

The second newsletter Rinaldi enclosed was dated 14 November 1988. It read:

"My longest and most troubled stay in Italy will come to a close in just about a month. Here is a brief summary on how things stand at the moment. No one expected people to react so strongly against the results of the carbon-14 test on the Shroud, and, too, against the stand of the Church, as expressed by Cardinal Ballestrero, regarding those results, not only in Italy but all over Europe.

"Cardinal Ballestrero has had reasons to regret what he said at the press conference on 13 October [that he saw no reason for the Church to doubt these results]. Briefly he stated that we must now revere the Shroud simply as an 'icon'. Sadly he made the point that, since the Church had trusted the scientists who had recommended and effected the carbon-14 test, it had no choice but to accept the verdict. When questioned at the conference whether a new carbon-14 test might be justifiable, his reply was a terse 'no'."

Rinaldi claimed that people in both religious and scientific circles were up in arms because of mistakes in test procedures and "the unprofessional behaviour of the scientists involved in the test". He said it was unclear what would happen next—the archdiocese will have a new head within the year, and Gonella was pressing for more tests by STURP. To the latter Ballestrero replied, "Let's not be in a hurry. I must see the Pope first".

Rinaldi pointed out that the *L'Osservatore Romano* had been strangely silent, publishing only the 13 October press communique. He interpreted this to mean that the Pope himself was not too happy with the way the entire matter of the carbon-14 tests were handled.

I wondered exactly what it was about the test procedures that might have made the pope unhappy. If they were the same ones that had initially made me and others unhappy, the pope had only himself to blame. It was he who cut Professor Chagas and the Pontifical Academy of Sciences out of the action and handed control to Luigi Gonella. Only the fact that the scientists in the three laboratories had conducted their measurements in such a professional manner and had independently reached agreement on the dates of the four pieces of cloth they had been given, saved the situation from disaster. Gonella had been damn lucky. Rinaldi's charge that the scientists' behaviour had been 'unprofessional' was quite preposterous. All along it was the behaviour of Gonella and his STURP cronies that had been unprofessional. I again thanked providence for ensuring that the carbon-14 results had not been tainted by any involvement by STURP. If the pope, by continuing to exercise benign neglect, permitted the shroud to suffer further assaults by STURP, so be it. At least they would not be desecrating a relic.

The 26 November 1988 issue of *Science News* carried some letters responding to their report of 8 October 1988 titled 'Shroud of Turin is a fake, official confirms'. One was by Paul Maloney of ASSIST (an organization formed to give support to STURP) who repeated the predictable arguments that the samples measured came from the same place on the shroud and could have been contaminated by handling, re-weaving or by damage of one sort or another in the 1532 fire. Another letter writer claimed no artist of the 14th century could have achieved the ångström accuracy (one ångström equals one part in 10 billion of a metre) in the dorsal/ventral symmetry of the image. I have not the faintest idea what he meant by this but it is *prima facia* nonsensical. A third letter writer pointed out that there was no evidence of pigment and suggested that: "One might also give thought to the possibility that whatever energetic process produced the flash photolytic 'burn' in the cloth might have dramatically altered both the atomic and molecular structure of the linen, thus resulting in faulty or artifactual carbon-14 readings". There is, of course, no scientific answer that can be given to people who invoke miraculous explanations.

Donahue phoned me on 23 January to say he had the videotape of the BBC programme produced by Neil Cameron. Doug claimed that there was a scene involving Woelfli where he says "but it's too young". I must say I have studied this videotape quite carefully and cannot pick out that phrase.

Ben Wiech sent me a note dated 6 February 1989 enclosing a couple of articles that had appeared in the British Catholic publication *The Tablet* of 14 January 1989. The first article, headed 'The future of the Shroud of Turin', noted that "some have been converted from unbelief by gazing on the remarkable face [of the shroud image] ... It is natural that those committed to its authenticity for a variety of reasons should seek to evade the verdict of the carbon testing ... [and] who in the teeth of the scientists' highly impartial and objective results will continue to deny the evidence because they have too much at stake. That is understandable, but those who have an open mind will be persuaded that the status of the shroud must be reassessed... There are some who feel the shroud's downgrading is a blow against belief: that a light has gone out and an assurance been lost that the finger of God can be seen at work in the world. Surely not. It was the scientists who began to claim too much for the shroud, and to suggest that it provided evidence for the Resurrection of a sort that could be tested in laboratories; now it is the scientists who have shown that such ideas are mistaken. There can indeed be no such evidence." (The only scientists who suggested the shroud provided evidence for the resurrection belonged to the STURP organization and they were true believers from the beginning. The carbon dating scientists involved never made any claims about the shroud's origin.)

The article noted that shortly before he died, Monsignor Jose Cottino, advisor on the shroud both to Cardinal Pellegrino and his successor, Ballestrero, remarked that he had never believed the relic was authentic because it would prove too much.

The second article was a question and answer interview with Teddy Hall by John Cornwall. Hall was asked why the measurements had not been performed blind as had been recommended at the Turin workshop. He said the problem was that the workshop had also recommended that the samples not be unravelled and the shroud weave was so distinctive its identity could not be concealed. He added that the Oxford tests were blind! After the samples were burned to carbon dioxide gas they were recoded by somebody separate from the carbon dating team, somebody who was sworn to secrecy. Thus neither Hall nor Hedges, who actually made the measurement, knew which of the four gas samples was the shroud. Hall said the other two labs did not take this precaution (he was wrong about that because in the paper that was published in *Nature* on 16 February 1989 it was stated that Zurich followed the same procedure as did Oxford). He made it clear that he thought this was a very wise and clever thing they had done at Oxford.

My personal view was that it had been a very silly and unwise thing to do. It meant that the two senior scientists at the Oxford AMS facility did not know whether they were actually measuring the samples Hall had so dramatically announced he had brought back to England from Turin. The same situation apparently also applied at Zurich. Whatever it was they measured at Oxford and Zurich fortunately bore a one to one relationship to the samples measured at Arizona where the fate of the samples was followed by Donahue from the cradle to the grave so to speak. It is not too outrageous to argue that the shroud sample was only measured at one laboratory—Arizona. The other two, as far as the senior scientists knew, measured a sample of carbon dioxide gas that just happened to give a carbon date close to the shroud's known historic date. I obviously do not believe this, but it was risky to have carried out the measurements in such a way. I am sure that Hall, Hedges and Woelfli would have argued that you had to trust someone. That was certainly true, but the people I would be most inclined to trust were the senior scientists at the three labs because they had the most to lose by improper behaviour. In the case of two of the labs, it could be argued that the senior scientists did not exercise a close enough control to merit such trust.

The Tablet article raised several other points with Hall that he answered effectively and satisfactorily. He was asked how he thought the image was produced. He replied he personally thought it was the result of a scorch either done cleverly with a poker or as a bas relief using a heated statue. As for the blood, he said, who knew whether it was human or pig's blood? He made it clear he was totally uninterested in resolving the question of how the image was produced.

Concerning the question of collaboration he had this to say: "There was absolutely no collaboration and we were very careful about this. As for being dispassionate, one of the scientific team in America, Doug Donahue, is a good Catholic and he believed firmly that it would show a date of the time of Christ. When the results came in he was very sad. I certainly didn't believe it was Christ's shroud. I don't deny that."

One of Hall's quotes that would raise a few eyebrows and might have offended people like Ben Wiech was "My own view is that many people don't want to know the truth about the Turin Shroud, and one is confronted with an extraordinary spectacle of mass auto-suggestion. People will persuade themselves that they can see all sorts of things in the image, because they are looking for such things in the first place." It was hardly a very diplomatic remark but it had a certain ring of truth to it, especially as concerned some of the people in STURP.

On 8 February 1989 I had a very interesting discussion with

Dick Kaeuper, a professor in the University of Rochester's History Department. He was writing a historical introduction to a book Geoffrey de Charny wrote on chivalry. The book was being translated for the first time from the original old French by Elsbeth Kennedy, a scholar at St Hilda's College, Oxford University. An illuminated copy of the manuscript existed in the Bibliotek Royale in Brussels, Belgium and a few US libraries had copies. Dick was interested in the shroud dating because its earliest recorded history dates to 1353–1356 AD when de Charny allegedly placed it in a church he had had constructed to house it in Lirey (near Troyes), France. According to Kaeuper, de Charny was an exemplar of a knight—literate and a rigid believer in chivalry. He had been imprisoned in England after being betrayed to the English by an Italian knight. When he returned to France after his release he had sought out his betrayer and caught him in the knight's castle in bed with his (the Italian's) English mistress. De Charny cut the man's head off on the spot, chopped his body in four pieces and mounted them around the castle walls. He then departed without further ado.

I lent Dick my copy of Ian Wilson's first book on the shroud and said I would also send him a copy of the talk I had given in Dubrovnik and my version of AMS carbon dating for laymen. A few months later, Kaeuper phoned me and said it was by no means absolutely established that de Charny was the original owner of the shroud or that the chapel he built was designed to house it. Even if de Charny were the shroud's owner in the mid 14th century, he may not have considered it overly noteworthy. In any case he never mentioned it in his book on chivalry. Dick described de Charny as playing the same role to aspiring knights as a coach does to members of his team. 'Look boys don't disport with harlots, etc.'

On 9 February 1989, Vernon Miller, the photographer and member of the STURP team that took such excellent photographs of the shroud during the 1978 tests, gave a talk on the shroud at the Rochester Institute of Technology, an institution which is well known for its photographic department. Shirley and I attended the talk. He had started by saying it was the first talk he had given since the carbon dating results had been announced. As I had predicted, he said that the dating was flawed because all the samples had been taken from the same place on the shroud. He claimed that linen fibrils are hollow 'like bamboo' and could be filled with smoke or water from the fire in 1532 which could not be removed by the cleaning processes used. I commented after his talk that the results were credible to me and it could hardly be a coincidence that the shroud's age corresponded to the historic one. It could readily be calculated that if the linen were really 1988 years old there would have to be 79% of the carbon in the linen that came from the fire and thus dated to the year 1532 for the combination to produce a 1325 AD date. The linen fibrils would have to be awfully hollow to contain that much carbon that was contemporary in 1532 to produce a date of 1352.

I met with him for about an hour at his hotel a couple of days later and he told me the 1 cm wide by 7 cm long strip that was cut from the shroud at the edge was just above where the Raes sample was taken. It still had stitches from where it had been sewn to the backing cloth. He said the 6 cm side strip which runs the length of the shroud was actually part of the main body of the shroud. Miller said the STURP people determined that a tuck was taken in the shroud which made it appear that the side strip is a separate piece of cloth. They speculated that this was done to strengthen the cloth for hanging it horizontally. Some 15 cm of this side strip was removed at some unrecorded time so only the basket weave backing cloth adjoins the place where both the Raes' sample and this 7 cm piece was removed. There was no question that both these samples were taken from the main body of the shroud and that there was no charred cloth anywhere near. Miller said that Riggi was writing a book that criticizes Gonella for selecting just one site for sampling. He also said that Meacham continued to maintain that the Raes sample was dated in California some time ago and was found to date to the first or second century. I assured Miller that even now (early 1989) there was no carbon dating facility in California capable of dating such a small amount of material and there never had been.

On 17 February 1989 I phoned Donahue. I asked him to send me a copy of the *Timewatch* programme and he said he would. Ted Litherland and I had been asked to organize a workshop on AMS at the *International Conference on Ion Beam Analysis* which was to be held at Queen's University—my old *Alma Mater*—in Kingston, Ontario, Canada in June 1989. One of the speakers I had suggested was Donahue, so in this phone call I asked him to send me a brief abstract of what he was proposing to say.

I told him that I had learned the shroud article would appear in the 16th February edition of *Nature*. Doug confirmed this and said that in this same issue was a letter from Thomas J Phillips of

Harvard and Fermilab speculating that a burst of neutrons at the instant of resurrection produced an excess of carbon-14. Such a neutron burst would also produce excess chlorine-36 if there were any chloride present on the cloth. He jokingly wondered whether Gonella would give Rochester a sample of the shroud to make a measurement of this rare radioisotope of chlorine. We were one of the few AMS labs that had a high enough energy tandem accelerator to make such a measurement and we had years of experience in doing so. Donahue said he had been contacted by the Tucson press about this letter by Phillips and it was the first time he had used expletives in public! He was indignant that *Nature* would publish such a letter.

I asked one of my high-energy colleagues which Harvard group was working at Fermilab. He said there was a huge collaboration on the colliding detector facility (CDF) working there that included Harvard. I called the Harvard Physics Department and the woman I talked to gave me Phillips' Fermilab number. She expressed surprise that Phillips would write such a letter. I finally phoned the place he was staying in Chicago—he was suffering from a touch of flu.

He seemed quite serious. I suggested he should try to give some plausible scientific explanation of how a biological system could produce neutrons. He said (as I later found he had in his *Nature* letter) that resurrections had never been studied scientifically. One cannot quarrel with that statement. I asked him whether he did not think it remarkable that, just by chance, the samples had been taken at just the right distance to change the date from 0 AD to 1325 AD, very close to the shroud's historic age. That did not seem to bother him. I noted that samples taken much nearer to the image (where, if his speculation were correct, the carbon-14 content would be even higher) would have given a carbon date even more recent than the historic age. I said that the only reason *Nature* would publish such a letter as his was because the writer was associated with such important institutions as Harvard and Fermilab.

I still had not seen the 16th February issue of *Nature* but was beginning to get news reports of the article and of Phillips' speculations. The 16 February 1989 edition of *USA Today* carried a story by Tim Friend and included some quotes from Paul Damon of Arizona: "These results provide conclusive evidence that the linen of the Shroud of Turin is medieval. This [speculation of people like Phillips] could go on *ad infinitum*. First, it is not good theology to place your faith in graven images. And scientifically we should strive to be as objective as possible and not introduce extra-scientific factors into our conclusions."

On 27th February the 16 February 1989 issue of the British journal *Nature* (volume 337) finally reached the library in my lab. On pages 611–615 appeared the article titled 'Radiocarbon dating of the Shroud of Turin' by P E Damon *et al.* As reported by Archbishop Ballestrero on 13 October 1988 it stated that there was a 95% probability that the flax from which the shroud's linen was made was harvested between 1260 and 1390 AD. The article was rather opaquely written—difficult to comprehend in complete detail even by experts in the field, but the results were clear. All three AMS carbon dating laboratories had independently obtained excellent agreement on the ages of the three control samples and the shroud sample.

The same issue of *Nature* carried the letter from T J Phillips, High Energy Laboratory, Harvard University, and a letter in reply (solicited by *Nature*) from Robert Hedges of Oxford. The two letters were headlined 'Shroud irradiated with neutrons?' Phillips' letter opened with, "If the shroud of Turin is in fact the burial cloth of Christ, contrary to its recent carbon-dated age of about 670 years, then according to the Bible it was present at a unique physical event: the resurrection of a dead body. Unfortunately this event is not accessible to direct scientific scrutiny..." Phillips then went on to speculate that, since the image on the shroud resembles a scorch, the resurrecting body must have radiated light and/or heat. Why then could it not also have radiated neutrons? (Why not indeed! In the world of miracles anything can happen.) The neutrons could have been captured by carbon-13 (a stable isotope present with all the carbon in the shroud) to form carbon-14. This extra production of carbon-14 (radiocarbon) could give the shroud a much later radiocarbon age. He pointed out that this extra radiocarbon would vary in amount from place to place on the shroud. Presumably it would be greater closer to the image where this postulated production of neutrons occurred. He added that the neutrons would also produce two other rare radioactive isotopes, chlorine-36 and calcium-41, and suggested an AMS search for these would settle the question.

Hedges reply was that Phillips proposed no plausible physical mechanism for producing such neutrons. If a supernatural explanation was proposed, then it was pointless to make any scientific measurements on the shroud. But the most devastating

NATURE VOL. 337 16 FEBRUARY 1989
————ARTICLES————
611

Radiocarbon dating of the Shroud of Turin

P. E. Damon[*], D. J. Donahue[*], B. H. Gore[*], A. L. Hatheway[*], A. J. T. Jull[*],
T. W. Linick[*], P. J. Sercel[*], L. J. Toolin[*], C. R. Bronk[‡], E. T. Hall[‡],
R. E. M. Hedges[‡], R. Housley[‡], I. A. Law[‡], C. Perry[‡], G. Bonani[§], S. Trumbore[‖*],
W. Woelfli[§], J. C. Ambers[¶], S. G. E. Bowman[¶], M. N. Leese[¶] & M. S. Tite[¶]

[*] Department of Geosciences, † Department of Physics, University of Arizona, Tucson, Arizona 85721, USA
‡ Research Laboratory for Archaeology and History of Art, University of Oxford, Oxford, OX1 3QJ, UK
§ Institut für Mittelenergiephysik, ETH-Hönggerberg, CH-8093 Zürich, Switzerland
‖ Lamont-Doherty Geological Observatory, Columbia University, Palisades, New York 10964, USA
¶ Research Laboratory, British Museum, London, WC1B 3DG, UK

*Very small samples from the Shroud of Turin have been dated by accelerator mass spectrometry in laboratories at Arizona,
Oxford and Zurich. As controls, three samples whose ages had been determined independently were also dated. The results
provide conclusive evidence that the linen of the Shroud of Turin is mediaeval.*

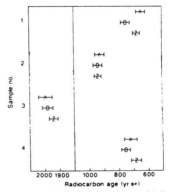

Fig. 1 Mean radiocarbon dates, with ±1σ errors, of the Shroud
of Turin and control samples, as supplied by the three laboratories
(A, Arizona; O, Oxford; Z, Zurich) (See also Table 2.) The shroud
is sample 1, and the three controls are samples 2-4. Note the break
in age scale. Ages are given in yr BP (years before 1950). The age
of the shroud is obtained as AD 1260-1390, with at least 95%
confidence.

*Article in Nature 16 February 1989 'Radiocarbon dating of the Shroud of Turin'.
The shroud dates to 1260–1390 AD at a 95% confidence level. (Reprinted with
permission from Nature **337** (1989) 302, © 1989 Macmillan Magazines Ltd,
and from P E Damon and D J Donahue.)*

argument against Phillips' idea was the fact that the samples were
taken at just the right spot on the shroud to produce its historic
date. A sample taken closer to the image would have produced
an even more modern date—even a date into the future!

No scientist has taken Phillips' neutron irradiation idea
seriously, but the possibility of contamination is constantly raised.
How much organic carbon contamination was required to change
0 AD to 1325 AD? The answer, mentioned previously in this
chapter, was that if the contamination occurred as a result of the

fire in 1532, then 79% of the shroud would have been composed of such carbon contamination and only 21% would have been actual carbon from the shroud linen. Such a possibility is preposterous, as anyone viewing the shroud samples before they were cleaned can attest. The other question that has been asked is: if the statistical probability that the shroud dates between 1260 and 1390 is 95%, what is the probability that it could date to the first century? The answer is about one in a thousand trillion, i.e. vanishingly small.

I received a letter dated 17 March 1989 from Phillips. It was clear that he was being serious about the neutron irradiation possibility. He said that as a high-energy physicist, he considered anything to be possible unless there was a reliable measurement that contradicted the possibility. He was surprised at Hedges' and my reaction to his suggestion of the possibility of a neutron flux accompanying Christ's resurrection. I basically disagreed with Phillips' approach to science, namely that anything was possible unless proved otherwise. The approach most scientists would take was to fit the data with the simplest possible set of assumptions. The most credible set of data on the shroud's age at present came from the radiocarbon measurements and they told us it dates to 1325 AD give or take some 30 years. Since the first record of the shroud dates to circa 1352 AD, most scientists would say that settled the question.

Some time in April 1989, Harbottle phoned from Brookhaven and asked if I had seen the article by Teddy Hall in the British journal *Archaeometry*. I had not since I am not an archaeologist and generally neither read nor am familiar with journals in that field. Harbottle expressed considerable annoyance at the general tone of the article. He said Hall painted himself as the man who was single-handedly responsible for bringing the Shroud of Turin to the test of carbon dating by AMS. His laboratory, along with those at Zurich and Arizona, had exposed the shroud as a fake dating only to the 14th century. Harbottle thought that some reply should be made to Hall.

I finally tracked down a copy of the article. It was published in the *Bulletin of the Research Laboratory for Archaeology and the History of Art—Oxford University*. The managing editors were M J Aitken and E T Hall of that laboratory. It was titled 'The Turin Shroud: an Editorial Postscript' by E T Hall and was a surprisingly biased account of the whole shroud adventure. After reading the article I was as annoyed as Harbottle and felt strongly that it must be answered. The straightforward way would be to write a letter to the editors of the journal but with Teddy Hall as one of the editors I figured there would be a fat chance of it being published.

Undaunted by my suspicion that anything I wrote in reply would be rejected by the journal's editors, I composed a letter and sent it off to the journal with a letter of transmittal dated 10 April 1989 as follows:

"Dear Sirs: I enclose an article in response to one by Professor E T Hall 'The Turin Shroud: An Editorial Postscript' *Archaeometry* **31** (1989) 92 which I request that you publish either as a letter to the editor or as a regular communication. I am sure that Professor Hall's sense of fair play will ensure that it will be published. Yours sincerely, H E Gove, Professor of Physics."

My article was titled 'Letter to the Editor: the Turin Shroud'. Hall's article and my reply are too long and detailed to be presented here. Readers who are interested in them can request copies from me.

On 17 April 1989, Aitken phoned acknowledging receipt of my article and said it would be published with some minor changes I readily accepted. I told him that I was very much obliged to him for being willing to publish it. He said he thought it was good to be able to present my viewpoint along with Teddy's. I must say, as a result, I recovered some of the esteem I had for Hall.

The Sunday 11 June 1989 edition of the *New York Times* carried a story headlined 'Despite Tests, Turin Shroud is Still Revered' written by Clyde Haberman. A summary of the points covered in the article follows:

• After eight months since radiocarbon tests showed it to be inauthentic, the shroud retains its allure for believers that it is Christ's shroud, according to the Reverend Felice Cavaglia, pastor of the Turin Cathedral where the object is stored.
• A group of specialists who met in Bologna in May 1989 concluded that contamination of the cloth over the years, including damage from a fire in 1532, could have undermined the tests.
• Harvard University physicist Thomas J Phillips suggested radiation during the resurrection could have altered the cloth's nuclear composition.
• A French monk charges that the British Museum official who coordinated the radiocarbon measurements substituted snippets of 14th century cloth for the postage-stamp-sized shroud samples. The charge has been denied. (The French monk in question was

Brother Bruno Bonnet-Eymard of whom I have more to say later.)
• Luigi Gonella, scientific advisor to the archbishop of Turin, stated he "had no scientific reason to think the testing was inaccurate". He expressed irritation that a premature leak indicated the shroud was a hoax. He said that conclusion was unwarranted but added that "even the law of gravity may turn out tomorrow to be in error".
• Writer Ian Wilson, who was converted to Catholicism by his studies of the shroud, stated "until somebody can show me how the image was made...the carbon dating, for reasons we don't know, may be in error".
• Pope John Paul II reiterated his often-stated conviction that the shroud "is certainly a relic" and not merely an icon. He told reporters recently "If it weren't a relic, one could not understand these reactions of faith that surround it and that express themselves even more strongly because of the scientific results".

Meyer Rubin sent me a copy of this article with the note: "Harry—I guess this was predictable. Don't trust those French monks. Best, Meyer."

I particularly liked the law of gravity quote from Luigi Gonella. Someone showed me a cartoon by S Harris. It was captioned 'The Shroud of London bears the likeness of Isaac Newton and has a tendency to fall to the ground'. It showed a table piled with books and papers off which has fallen a piece of cloth shown lying on the floor on which Newton's image appears. I added part of the clipping from the *New York Times* which read: " ' I have no scientific reason to think the testing was inaccurate,' said Luigi Gonella, scientific advisor to the Turin Archdiocese. . . .'Any scientific result is acceptable on its face value,' he said. 'Yet even the law of gravity may turn out tomorrow to be in error'." I wrote a note on it dated 21 July 1989: "Dear Luigi: How incredibly perceptive you are! To equate the probability that the shroud date is wrong with the probability that the law of gravity is in error requires a razor sharp mind. I could not agree more. Best regards, Harry." I never sent it to Gonella—I suppose I felt I had done enough to alienate him.

In April 1991, a glossy booklet appeared that claimed the dating of the shroud to medieval times was a scientific forgery. It was widely circulated *gratis* to everyone even remotely involved in the shroud carbon dating. The publication in question was the combined March and April 1991 monthly editions, in English, of the *Catholic Counter-Reformation in the XXth Century* Nos 237 and 238. The ultra-conservative Catholic organization (headquartered in France) responsible for this publication opposes the Vatican on a number of issues. In particular, it is firmly convinced that a grand conspiracy involving Tite, Ballestrero, Gonella, and the scientists in the three laboratories produced a false medieval date for the Turin Shroud and the Counter-Reformation organization has spent a considerable amount of money to try to prove its case. A member of this organization, Brother Bruno Bonnet-Eymard, visited Tucson, Zurich, Oxford and the British Museum in London in his attempt to obtain evidence that the medieval date was a hoax. Most knowledgeable people receiving this publication read it with a combination of amusement and contempt. None but members of the right-wing zealots comprising this Catholic splinter group takes their fraud claims seriously.

They did, however, make a good case for the claim that more than the 1 cm × 7 cm cloth sample was removed from the shroud. Some part of this extra material was later sent by Turin to the IsoTrace AMS carbon dating laboratory run by Ted Litherland at the University of Toronto. It was not identified by the customer as coming from the Turin Shroud but the scientists at Toronto were readily able to identify it from its unique weave. The measurement at IsoTrace agreed with the values Arizona, Oxford and Zurich published in *Nature*. As far as Toronto was concerned it was just another measurement for which they charged their usual fee and they made no claim as to its identity.

In 1992 Holger Kersten and Elmar Gruber published a book in German, later translated into English, titled *The Jesus Conspiracy*. It was published in 1994 in Great Britain, Australia and the USA. The authors attempt to make the case that Christ did not die on the cross but was resuscitated in the burial tomb to which he had been taken after being removed from the cross. As soon as he was able to walk he was assisted out of the tomb to some unspecified destination. The linen cloth on which he had been laid and which, for some reason, had been folded over his head to cover the front of his body had imprinted on it, by an unexplained process, his frontal and dorsal image. The authors claim that this image discloses many examples of evidence that it is not that of a corpse but of a living badly wounded man. They allege that, at the time of sample removal for carbon dating in April 1988, somehow 14th century cloth identical in weave to that of the Turin Shroud was substituted for it. It was this substitute that Michael Tite inserted, along with the three control samples, into the stainless steel containers under the watchful gaze of Cardinal Ballestrero and Professor Luigi Gonella. This, of course, is a parroting of the allegations made by Brother Bruno Bonnet-Eymard mentioned above. Kersten and Gruber's ideas seem fanciful to me but they evidently sell books.

In 1994 another book titled simply *Turin Shroud* by Lynn Picknett and Clive Prince was published. The authors say they accept the carbon-14 date of 1325 AD for the shroud's linen but argue that the double image on the shroud was created by Leonardo da Vinci around 1492 using a photographic process he invented. When, in a letter to the authors, I pointed out that the probability the cloth that was carbon dated to the year 1325 could actually have been young enough to have been the cloth that Leonardo used in 1492 was only one in thirty thousand they replied that Leonardo would scarcely have used 'linen fresh from the loom'. If one accepts that, Leonardo would have had to have been prescient enough to recognize that the age of cloth could be scientifically determined and thus to choose linen that was some 150 years old (to correspond to the shroud's historic date). That is to say, not only did he invent photography but also predicted a cloth dating technique (did Willard Libby's Nobel prize go to the wrong person?). Furthermore, what happened to the Lirey shroud? All this strikes me as a delightful fantasy.

In 1995 at the annual meeting of the American Chemical Society, a group of Russian workers led by Dr Dimitri Kouznetsov presented a paper titled 'The shroud of Turin: a progress report on research into the old textile radiocarbon dating results'. They listed as their home institution the E A Sedov Biopolymer Research Laboratories in Moscow. They also submitted a paper to the *Journal of Archaeological Science*. In both the talk and paper they claimed that they had shown experimentally that a linen textile, a little over 2000 years old, when heated in air to 200 °C for 90 minutes was carbon dated by AMS to be 1400 years younger. They argued that this duplicated the conditions of the 1532 Chambery fire and meant the measured carbon date of 1325 AD published in *Nature* by Damon *et al* was older, *mirabile dictu*, by approximately that many years. In an accompanying paper in the *Journal of Archaeological Science* A J T Jull, D J Donahue and P E Damon—the three senior members of the AMS facility at the University of Arizona who carried out the first carbon dating of the shroud—disputed the Kouznetsov *et al* claim. They duplicated the experiments carried out by the Russian group and concluded that "the attack by Kouznetsov and his co-workers on measurements of the radiocarbon age of the Shroud of Turin and on radiocarbon measurements on linen textiles in general are unsubstantiated and incorrect. We also conclude that other aspects of the experiment are unverifiable and irreproducible." That is about as strong language as reputable scientists generally permit themselves to use when disputing what they regard as a fraudulent scientific claim.

There is, however, one other development that should be taken seriously. All the reasons, mentioned above, that have been advanced for the age of the shroud being younger than 1325 years, range from the highly improbable to the ludicrous. This one alone, at least so far, merits further detailed investigation.

On 2 and 3 September 1994 a *Round Table on the Microbiology of Ancient Artifacts* was held in San Antonio, Texas. It was organized by Dr Stephen Mattingly and Dr Leoncio A Garza-Valdes of the Department of Microbiology of the University of Texas Health Science Center at San Antonio (UTHSCSA). Garza-Valdes is also a medical doctor with a pediatric practice in San Antonio, Texas. The main subject of the round table was a discussion of bio-plastic coatings produced by bacteria and fungi and found on the surface of ancient artifacts, desert rocks (where it is referred to as desert varnish) and around the fibres of some ancient textiles including the Turin Shroud. These bacteria take their nourishment from the air and hence can be adding carbon with a component of carbon-14 contemporaneous with the time of bacterial growth— and thus younger than the cellulose of the ancient textile. With the assistance of Giovanni Riggi (the person who removed a sample from the shroud in 1988 for radiocarbon dating by AMS) Garza-Valdes obtained a small sample of cloth from the shroud taken from the same area as those used in the AMS measurements. Microscopic examination showed a definite halo or bio-plastic coating of varying thickness around the fibres. The UTHSCSA researchers established that the acid–base–acid cleaning method employed on shroud samples by the three AMS carbon dating laboratories left the bio-plastic coating intact. They conclude that this coating added carbon-14 to the shroud cellulose thus causing the AMS carbon date to be too young. How much too young has not yet been established and, indeed, will not be easy to establish. Research on the question is continuing.

In September 1995, Giovanni Cardinal Saldarini, Archbishop of Turin and Custodian of the Holy Shroud, who succeeded to the post when Cardinal Ballestrero retired, issued a 'Declaration on the Experiments in Regards to the Holy Shroud'. In summary it states that no new pieces of material have been removed from the shroud since 21 April 1988 and no residue of the material removed then remains; if it does no permission has been given by the shroud's custodian to have it or to use it and it must therefore be returned, the results of any experiments performed on such material have no value since it cannot be certain true shroud material was used. This does not refer to the research carried out in October 1978 by STURP *et al* and the carbon dating carried out by the three AMS laboratories and, finally, the Holy See and the Archbishop of Turin invites the scientists to be patient until a clearer research programme can be orchestrated.

For the time being, at least, the adventure is over—it lasted too long and was filled with too much acrimony. I felt no joy in the final result except that it proved the power of AMS to credibly date precious artifacts. Probably never again would any of the carbon-14 participants let themselves get involved in dating any religious artifact or relic except as a strictly commercial venture.

What is the answer to the question as to whether the Turin Shroud is a relic, an icon or a hoax? In my view it is certainly not a hoax and, unless a plausible, scientifically valid reason is found for the radiocarbon date being too young, it cannot be a relic. I believe it is an icon and, arguably, the most important icon in Christendom at that.

Meanwhile, the shroud still lies in its silver cask in the Cathedral of John the Baptist in Turin. A new archbishop presides over the Turin Archdiocese. No further scientific forays on its most famous artifact have been authorized. The mystery of how its image was produced remains just that—a mystery.

Maybe that is as it should be.

Cast of Characters

Adler, A D (Alan): Professor of Chemistry at Western Connecticut State University. An expert on blood chemistry and a member of the Shroud of Turin Research Project. He attended the workshop on dating the Turin Shroud.

Allen, K W (Ken): Professor of Physics (nuclear experimental) at Oxford University. Head of the university's nuclear physics accelerator laboratory.

Apers, D: Professor in the Laboratory of Inorganic and Nuclear Chemistry, University of Louvain.

Ballestrero, A (Anastasio): Cardinal and Archbishop of Turin. Designated by the pope as custodian of the Turin Shroud. He served as patron of the workshop on dating the Turin Shroud. Sometimes referred to as the bishop.

Beasley, T E (Tom): A geochemist at the Department of Energy in Argonne National Laboratory and presently at their Environmental Measurement Laboratory in New York.

Beukens, R P (Roelf): Involved in the early work on accelerator mass spectrometry (AMS) at the University of Rochester. Research Scientist at the IsoTrace AMS Laboratory at Toronto.

Bohr, A (Aage): Professor of Physics (theoretical nuclear) at the Niels Bohr Institute in Copenhagen. Nobel Laureate in Physics in 1975. Member of the Pontifical Academy of Sciences.

Bollone, P L B (Baima): Member of the Institute of Forensic Medicine of the University of Turin. For many years he researched possible mechanisms for the formation of the image on the Turin Shroud.

Bonisteel, R (Roy): Host of the Canadian Broadcasting Corporation's TV programme *Man Alive*.

Bonnet-Eymard, Brother Bruno: Member of the Catholic Counter-Reformation in the XXth Century group. Believes there was a conspiracy to produce a medieval date for the Turin Shroud.

Bowman, S G E (Sheridan): A scientist in the British Museum's Research Laboratory. She presented the paper on the Museum's interlaboratory comparison tests at Trondheim and was a co-author of the paper on the results of the carbon dating of the shroud.

Brignall, S L (Shirley): Administrative Assistant to the chair of the Department of Physics and Astronomy, University of Rochester. She was involved in many aspects of planning to date the Turin Shroud. She attended the workshop on dating the shroud.

Burleigh, R (Richard): Scientist with the Research Laboratory of the British Museum. Co-author of the Museum's paper on the interlaboratory comparison test presented by Bowman at Trondheim.

Cameron, N (Neil): Producer of television programmes for the British Broadcasting Corporation (BBC). In particular he produced a programme on dating the Turin Shroud for the BBC's *Timewatch* series.

Canuto, V (Vittorio): Theoretical astrophysicist and senior scientist at the NASA Institute for Space Studies in New York. He served as a science advisor to Chagas and, along with Gove, assisted Chagas in organizing the workshop on dating the Turin Shroud. He participated in the workshop.

Casaroli, Cardinal: Vatican secretary of state.

Cavallo, G (Giorgio): Professor of Biology, University of Turin. President of the 1978 Congress on the Turin Shroud.

Celli, Monsignor: Vatican ambassador to the United Nations.

Chagas, C (Carlos): Professor and medical doctor, Institute of Biophysics, Federal University, Rio di Janeiro, Brazil, president of the Pontifical Academy of Sciences. He chaired the workshop on dating the Turin Shroud.

de Charny, Geoffroi I: French knight, first introduced the shroud to the world. He died a few years later in 1356 without revealing its origin.

Cheshire, L (Leonard): Bomber pilot in the Royal Air Force in WW II. Winner of the Victoria Cross in 1944. He was Churchill's observer of the atom bombing of Hiroshima. Past president of the British Turin Shroud Society. He founded the Cheshire Foundation for young disabled people. He died on 31 July 1992.

Clark, K R (Kenneth): Reporter for the Chicago *Tribune*.

Coero-Borga, P (Piero): The person in Turin to whom the

Rochester/Brookhaven offer to date the Turin Shroud was submitted with the request to pass it on to the archbishop of Turin. He never did so.

Conard, Nick: A graduate student at the University of Rochester.

Corretto: Washington correspondent for *La Stampa*.

Cottino, Monsignor Jose: A catholic priest in Turin. He served as press spokesman and caretaker of the shroud for Ballestrero.

D'Amato, A (Alphonse): Junior US senator from New York State.

Damon, P E (Paul): Professor, Department of Geosciences, University of Arizona and co-director of the Accelerator Mass Spectrometry Laboratory at Arizona. He attended the workshop on dating the Turin Shroud in 1986 and his was one of the three laboratories that dated the shroud in 1988.

Dardozzi, R (Renato): An electrical engineer and director of the Pontifical Academy of Sciences. He attended the workshop on the dating of the Turin Shroud.

Dinegar, R H (Bob): Research chemist at Los Alamos National Laboratory, New Mexico, an Episcopalian minister and an early member of the Shroud of Turin Research Project. He was chair of their radiocarbon dating committee. He attended the Turin workshop on dating the shroud.

D'Muhala, T (Thomas): President of the Nuclear Technology Corporation in Connecticut and a senior director of STURP.

Donahue, D J (Doug): Professor of Physics (nuclear experimental) at the University of Arizona and co-director of the Accelerator Mass Spectrometry Laboratory at Arizona. He attended the workshop on dating the Turin Shroud in 1986 and his was one of three laboratories that dated the shroud in 1988.

Druzik, J (James): An art conservator with the J P Getty Museum in California.

Duplessy, J-C (Jean-Claude): A senior scientist at the accelerator mass spectrometry facility at Gif-sur-Yvette, France. He attended the workshop on dating the Turin Shroud.

Dutton, D (Denis): Member of the School of Fine Arts, University of Canterbury, Christ Church, New Zealand.

Evin, M J (Jacques): Head of the radiocarbon laboratory, University of Claude Bernard, Lyon. He attended the workshop on dating the Turin Shroud.

Fleming, S (Stuart): A scientist working in Hall's laboratory in Oxford and later in the museum at the University of Pennsylvania.

Flury-Lemberg, M (Mechthild): Head of the Textile Workshop of the Abegg-Stiftung Institute in Switzerland. She was selected by Chagas to be present when samples were removed from the shroud. She attended the workshop on dating the Turin Shroud.

Frei, M (Max): A Swiss forensic scientist. He examined pollen samples from the shroud.

Galileo: Punished by the Catholic Church in the 17th century for arguing that the Earth moved around the Sun. Pope John Paul II apologized for this in 1979.

Gonella, L (Luigi): Professor of Physics (metrology) at the Turin Polytechnic Institute. He served as science advisor to the archbishop of Turin. He attended the workshop on dating the Turin Shroud.

Gove, Elizabeth A (Betty): Wife of author.

Gove, H E (Harry): Co-inventor of accelerator mass spectrometry along with Litherland and Purser. Professor of Physics (now emeritus) at the University of Rochester. Director of the Nuclear Structure Research Laboratory at the University of Rochester 1963–1988. He and Canuto assisted Chagas in organizing the workshop on dating the Turin Shroud. He attended the workshop and was present in Arizona when the shroud was first dated on 6 May 1988.

Hall, E T (Teddy): Professor and head (now retired) of the Research Laboratory for Archaeology and the History of Art, Oxford University, Oxford, England. This department includes the Radiocarbon Accelerator Unit. He attended the workshop on dating the Turin Shroud and his was one of the three laboratories involved in dating the shroud.

Harbottle, G (Garman): A senior scientist in the Chemistry Department of Brookhaven National Laboratory. He is in charge of a small counter radiocarbon dating laboratory. He attended the workshop on dating the Turin Shroud.

Hedges, R E M (Robert): A senior scientist at the Radiocarbon Accelerator Unit of the Research Laboratory for Archaeology and the History of Art, Oxford University. He attended the workshop on dating the Turin Shroud and his was one of the three laboratories involved in dating the shroud.

Heller, J (John): Medical doctor in New England. Blood expert.

Highfield, R (Roger): Reporter for the British newspaper, the *Daily Telegraph*.

Hill, S (Seth): Associated with the TV programme *In Search of* hosted by Lenard Nemoy.

Jackson, J P (John): Faculty member of the US Air Force Academy in Colorado Springs. Co-founder of the Shroud of Turin Research Project along with Jumper.

Jennings, P (Peter): Freelance writer and reporter especially on shroud matters.

Jull, A J T (Tim): One of the scientists at the University of Arizona's Accelerator Mass Spectrometry Laboratory. He was involved in the carbon dating measurements on the Turin Shroud.

Jumper, E J (Eric): Faculty member of the US Air Force Academy in Colorado Springs. Co-founder of the Shroud of Turin Research Project along with Jackson.

Kahle, B (Brian): Host of a programme on TV Channel 7, Buffalo, New York.

Kaeuper, R W (Richard): Professor of History, University of Rochester.

Krenis, L (Lee): Member of the Public Relations Department of the University of Rochester.

Leczinsky-Nylander, C (Christine): Producer of science and cultural programmes for the Swedish Broadcasting Corporation.

Leonardo da Vinci: 1452–1513, a renowned medieval artist.

Libby, W F (Bill): Nobel Laureate in Chemistry for 1960 for the development of the radiocarbon method of dating. He died in 1980.

Litherland, A E (Ted): Co-inventer of accelerator mass spectrometry (AMS) along with Gove and Purser. Professor of Physics at the University of Toronto. Director of the IsoTrace AMS laboratory at Toronto.

Luckett, R (Richard): A librarian at the Pepys Library, Magdalene College, University of Cambridge.

Lukasik, S (Stephen): Vice President–Technology, Northrup Corporation, Los Angeles. He was in charge of experiments the STURP group hoped to carry out on the shroud. He attended the workshop on dating the Turin Shroud.

Maloney, P (Paul): Director of the Association of Scholars International for the Shroud of Turin.

McClellan, W (William): Reporter for the St Louis *Post Dispatch* and son-in-law of Douglas Donahue.

McCrone, L (Lucy): Wife of Walter McCrone.

McCrone, W (Walter): President of McCrone Associates, a microscopy/chemistry laboratory in Chicago. On the basis of studies he made he claimed both the Vinland Map and the Turin Shroud were fakes.

Meacham, W (Bill): An archaeologist at the Centre of Asian Studies in Hong Kong. He attended the workshop on dating the Turin Shroud.

Miller, V (Vern): STURP photographer at the Brooks Institute, Santa Barbara.

Milton, J C D (Doug): Nuclear physicist at Atomic Energy of Canada, Ltd, Chalk River. A member of their AMS team.

Muller, R A (Rich): Professor of Physics at the University of California at Berkeley. He invented the idea of radiocarbon dating by accelerator mass spectrometry independently of and virtually simultaneously with the University of Rochester group.

Nelson, D E (Erle): Professor in the Department of Archaeology at Simon Fraser University in British Columbia. He was one of the early pioneers in the field of radiocarbon dating by accelerator mass spectrometry. He carried out the first measurements of carbon-14 in natural material virtually simultaneously with the group at the University of Rochester.

Nickell, J (Joe): Author, magician, detective. He believed the shroud was a fake and published his theories in several articles and a book.

Nydal, R (Reidar): Professor, University of Oslo and director of the university's carbon dating laboratory.

Oeschger, H (Hans): Director of the Physics Institute, University of Bern, one of the world's leading carbon dating laboratories employing the standard decay counting technique.

Otlet, R L (Bob): Head of the Isotope Measurement Laboratory at the Atomic Energy Research Establishment, Harwell. This laboratory has a small-counter radiocarbon decay counting component. He attended the workshop on dating the Turin Shroud.

Otterbein, Father: President of the Holy Shroud Guild in the USA.

Phillips, T J (Thomas): Researcher at Harvard University's high-energy physics laboratory, Cambridge and at Fermilab, Chicago.

Plummer, M (Mark): Executive director of the journal, *Skeptical Inquirer*, published in the USA.

Pope John Paul II: Pope of the Roman Catholic Church and owner of the Turin Shroud.

Purser, K H (Ken): Co-inventer of accelerator mass spectrometry along with Gove and Litherland. President of General Ionex Corporation (now Southern Cross Corporation). Chief accelerator physicist in charge of the installation of the MP tandem Van der Graaff at the University of Rochester.

Raes, G (Gilbert): Professor and textile expert, Ghent University.

Riggi, G (Giovanni): He was the person who cut the sample from the shroud for radiocarbon dating by accelerator mass spectrometry. He attended the workshop on dating the Turin Shroud.

Rinaldi, P (Peter): A catholic priest connected with the Cathedral in Turin and also a parish near New York City. He has been associated with the shroud since a choir boy in Turin. To many people he is 'Mr Shroud'. He is now deceased.

Robinson, J A T (John): Bishop of the Church of England, Dean of Chapel, University of Cambridge, member of the British Turin Shroud Society, author of *Honest to God*. He is now deceased.

Rogers, R (Ray): A chemist at Los Alamos National Laboratory and a member of STURP.

Rose, K (Kenneth): A reporter for the London *Sunday Telegraph*.

di Rovasenda, E (Enrico): A Catholic priest and past director of the Pontifical Academy of Sciences. He attended the workshop on dating the Turin Shroud.

Rubin, M (Meyer): A research geologist and head of the US Geological Survey's radiocarbon laboratory in Reston, Virginia. One of the early collaborators with the accelerator mass spectrometry group at the University of Rochester's Nuclear Structure Research Laboratory.

Sayre, E (Ed): Scientist with the Boston Museum of Fine Arts and a member of Harbottle's carbon dating group at Brookhaven National Laboratory.

Schmidt, F (Fred): Professor of Physics (nuclear experimental), University of Washington, Seattle. Started an AMS programme there. He is now deceased.

Smalley, K (Kathy): Director of the Canadian Broadcasting Corporation's television programme *Man Alive*. For the programme in this series on the Turin Shroud film was shot at the Nuclear Structure Research Laboratory in Rochester and in Turin during the 1978 conference on the shroud.

Somolo, Cardinal Martinus: Vatican undersecretary of state.

Sox, H D (David): Episcopalian priest and former secretary of the British Turin Shroud Society. He teaches at the American School in London. Author of several books on the shroud.

Stevenson, K (Ken): Public relations official for the Shroud of Turin Research Project (STURP) during the 1978 conference on the shroud that was followed by STURP's five day scientific investigation of properties of the shroud.

Stoenner, R W (Dutch): Colleague of Harbottle at the small-counter radiocarbon decay counting laboratory in Brookhaven National Laboratory.

Stuiver, M (Minze): Head of the carbon dating laboratory at the University of Washington, Seattle.

Sullivan, W (Walter): Senior science writer for the *New York Times*.

Testore, Professor Franco A: An Italian textile expert from the Turin Polytechnic's Department of Material Science. He was present when the sample was removed from the shroud for carbon dating.

Thomas, M (Michael): Religion correspondent of *Rolling Stone*.

Tite, M S (Mike): Head of the Research Laboratory of the British Museum. He attended the workshop on dating the Turin Shroud. He participated in the shroud sampling in Turin and he and two colleagues at the British Museum were involved in the dating of the shroud. He became Hall's successor at the Research Laboratory for Archaeology and the History of Art at Oxford University.

Umberto II: Former king of Italy. Died in 1983 in exile in Portugal.

Uznanski, H: Head of the National Science Foundation's Cooperative Science Program.

Vial, M G (Gabriel): A French textile expert from the Historical Museum of Fabrics, Lyon.

Weaver, K (Ken): Senior assistant editor of *National Geographic*.

Weisskopf, V F (Viki): Professor of Physics (nuclear and particle theory), Massachusetts Institute of Technology (now emeritus), and member of the Pontifical Academy of Sciences.

Wiech, B (Benjamin): A lawyer from Buffalo, New York.

Willard, H (Harvey): Head of the Nuclear Physics Program of the National Science Foundation, Washington, DC.

Wilson, Sir David: Director of the British Museum.

Wilson, I (Ian): British author of several books on the Turin Shroud and an official of the British Society for the Turin Shroud.

Wilson, W (William): US ambassador to the Vatican.

Woelfli, W (Willy): Director (now retired) of the accelerator mass spectrometry laboratory in Zurich, Switzerland. He attended the Turin workshop on dating the Turin Shroud in 1986 and his was one of three laboratories that ultimately dated the shroud in 1988.

Zugibe, F: A medical doctor who studied the effects of crucifixion.

Chronology

1353 It is thought that the Lirey shroud (now generally believed to be the same as the Turin Shroud) came into the possession of Geoffroi I de Charny, a French seigneur or knight, around this time. A church in which to store it was completed in 1356 in Lirey, France. It is not known how it came into de Charny's possession. He died in battle in 1356 and the secret of how he came to own the shroud died with him. It was exhibited in 1357 (the shroud's most secure date) amidst considerable controversy.

1389 Geoffroi II de Charny, son of Geoffroi I, held another exhibition of the Lirey shroud, again rousing controversy. He died in 1398 and his daughter Margaret became the shroud's owner

1453 Around this time the shroud, in some complicated way, passed from Margaret's hands and became the possession of the House of Savoy. In the sixty years between 1390 and 1450 details on the Lirey Shroud's whereabouts are remarkably vague.

1502 The shroud was moved to the chapel of a Savoyard church in Chambery, France.

1532 A fire broke out in the sacristy of the Chambery church and molten gouts of silver burned through the shroud.

1534 Poor Clare nuns cut away charred cloth around the burn holes and covered them with triangular pieces of cloth. They also applied a backing cloth to the shroud.

1578 The seat of the house of Savoy moved to Turin, Italy and the shroud moved with it.

1898 An eight day exposition of the shroud was held. Secondo Pia took the first photograph of the shroud and discovered that the full size dorsal and frontal images of a crucified man on it were negatives.

1939 At the outbreak of World War II the shroud was moved to the Benedictine monastery of Montevergine in the mountains of southern Italy. It was returned to Turin in 1946.

1969 The shroud was shown to members of a special scientific commission. New photographs in both colour and black and white were taken.

1973 The shroud was shown on TV. Max Frei, a Zurich forensic expert, removed pollen samples and two small pieces were cut from the main body of the shroud (5 square centimetres) and from the side strip (3 square centimetres) for study by Professor Gilbert Raes, director of Ghent University's textile laboratory.

1977 *May.* Accelerator mass spectrometry (AMS) was invented at the University of Rochester by a group of scientists from the General Ionex Corporation, the University of Rochester and the University of Toronto and independently at the Lawrence Berkeley National Laboratory and the first application of AMS for carbon-14 dating took place at the University of Rochester.

1977 *June.* The University of Rochester issued a press release on AMS and articles on the subject were published in *Time* magazine and the *New York Times*. As a result of the *Time* magazine story a letter was received at Rochester from the Reverend H David Sox asking whether the new technique could be used to date the Turin Shroud. The answer was yes but that it was a bit too soon to do so.

1978 *October.* The *Second International Congress on the Shroud*, Turin, Italy was held. A paper was submitted to the congress by the Rochester group and another by the Brookhaven group describing how the shroud could be carbon dated using very small samples of cloth. The shroud had been exhibited in the Cathedral of John the Baptist in Turin from 27 August to 8 October and this was followed by the international congress on 9 and 10 October and then by five days of 'non-destructive' tests by members of STURP and a few other scientists.

1979 *February.* A letter was mailed to the archbishop of Turin on behalf of the Rochester, Toronto, General Ionex, US Geological Survey group and the Brookhaven National

Laboratory group offering to date the Turin Shroud using milligram samples of cloth.

1979 *March.* A meeting was held in Santa Barbara, California to discuss the preliminary analysis of data obtained by STURP in October, 1978. At the meeting Walter McCrone claimed his studies of the shroud image suggested it was painted by an artist.

1982 At an archaeometry conference held in Bradford, England Robert Otlet of the Atomic Energy Research Establishment at Harwell, England suggested that the British Museum provide samples of cloth for carbon dating to those laboratories who had expressed a willingness to carbon date the Turin Shroud to test their ability to do so (this became the laboratory intercomparison test).

1983 *January.* Michael Tite of the British Museum's Research Laboratory informed the carbon dating labs at Arizona, Bern/Zurich, Brookhaven, Harwell, Oxford and Rochester that he was prepared to carry out the intercomparison tests Otlet had suggested. The six labs agreed to participate.

1985 *June.* The *12th International Radiocarbon Conference* is held in Trondheim, Norway. The British Museum presents the results of the laboratory intercomparison test. Harry Gove organizes a meeting of representatives of the six laboratories, the British Museum and a representative from STURP. It was suggested that the Pontifical Academy of Science be contacted and that Gove prepare a protocol for carbon dating the shroud.

1985 *August.* The final version of the protocol was completed and sent to the carbon dating laboratories, the British Museum and to the President of the Pontifical Academy of Science.

1985 *October.* Gove and Shirley Brignall of the University of Rochester met with Carlos Chagas, president of the Pontifical Academy of Science, and Vittorio Canuto, a NASA astrophysicist and a scientific aide to Chagas, at the Holy See Mission to the United Nations in New York to discuss the possibility of holding a workshop on dating the shroud involving all the interested parties.

1985 *November.* Canuto informed Gove that Chagas had received permission from the Vatican and the archdiocese of Turin to hold the workshop in June, 1986.

1986 *January.* Gove submitted a proposal to the National Science

Foundation's Cooperative Science Program with Western Europe to cover the expenses for three US carbon dating scientists, one from each of the US laboratories, Arizona, Brookhaven and Rochester, to attend the workshop.

1986 *February*. Gove and Shirley Brignall met with Professor Luigi Gonella, science advisor to the archbishop of Turin, at the Holy See Mission to the United Nations in New York to discuss the Turin workshop on carbon dating the shroud. Gonella insisted that the workshop be held at his institution, the Turin Polytechnic. Gove noted that the workshop is being organized by the Pontifical Academy of Science and should be held in their headquarters in the Vatican. Gonella opposed the idea that six carbon dating labs be involved.

1986 *April*. Chagas sent letters of invitation to attend a workshop on carbon dating the Turin Shroud to be held in the Archbishop's Palace in Turin 9–11 June 1986. A few days before doing so, however, he revealed to a British reporter, Peter Jennings, that the meeting would be held in June and an article by Jennings to this effect was published.

1986 *May*. The National Science Foundation approved funding for expenses of three US participants in the Turin workshop. A few days later the Pontifical Academy of Sciences announced that the June workshop had been postponed. It later turned out that this postponement was triggered by the article by Jennings. A cable was composed by Gove and sent to the archbishop of Turin, the Vatican secretary of state and the president of the pontifical academy. It was signed by representatives of four of the six carbon dating labs and the British museum and strongly protested the postponement of the workshop.

1986 *July*. Chagas sent a second invitation to attend the workshop on 29 September through 1 October 1986. It would be held in Turin in an institution associated with the archdiocese of Turin and not at Turin Polytechnic. Chagas invited Gove to prepare an agenda for the workshop.

1986 *August*. Shirley Brignall and Gove went to New York City to consult Canuto and Ambassador Celli on the agenda Gove had prepared. The agenda was mailed to Chagas and he eliminated the reference to 'six' laboratories and the revised agenda was mailed to all workshop invitees in mid-September.

1986 *September*. Brignall and Gove travelled to Rome, visited

Chagas in the Pontifical Academy of Science headquarters in the Vatican and were introduced to Pope John Paul II. The Turin workshop on carbon dating the Turin Shroud chaired by Chagas was held in a seminary in Turin from 29 September to 1 October 1986. A protocol for dating the shroud was agreed upon.

1987 *April.* The *Fourth International Symposium on Accelerator Mass Spectrometry* was held at Niagara-on-the-Lake, Ontario, Canada 27–30 April. It celebrated the tenth anniversary of the first detection of carbon-14 in natural organic material by accelerators. No word had come concerning a decision from Turin or Rome to date the shroud. People at the symposium representing the five AMS labs that had participated in the Turin workshop agreed to confirm to Chagas their support of the dating protocol and to press for action. During the symposium an article was published in the Turin newspaper *La Stampa* quoting Gonella as stating that only two or three carbon dating laboratories would be given shroud samples to date.

1987 *May.* A letter signed by representatives of the five AMS laboratories was mailed to Chagas. It confirmed their support for the carbon dating protocol and rejected the proposed changes reported by *La Stampa*.

1987 *June.* Conference phone call was made involving Canuto, Donahue (Arizona), Harbottle (Brookhaven) and Gove (Rochester). It became clear during the call that the number of labs to be involved in carbon dating the shroud would indeed be reduced to two or three as indicated in Gonella's interview with *La Stampa*.

1987 *July.* A cable was sent to Cardinal Ballestrero, Archbishop of Turin, with copies to Cardinal Casaroli, Vatican secretary of state, and Professor Chagas, president of the Pontifical Academy of Sciences protesting a reduction in the number of carbon-14 labs to be involved in dating the shroud from the original seven to two or three. It was signed by Donahue (Arizona), Duplessy (Gif-sur-Yvette), Gove (Rochester), Hall (Oxford), Harbottle (Brookhaven), Otlet (Harwell), Tite (British Museum) and Woelfli (Zurich).

1987 *August.* Hall informed Gove that, in the future, he intended to keep quiet and await developments from Italy.

1987 *October.* The archbishop of Turin informed all the participants in the Turin workshop that Arizona, Oxford and

Zurich had been selected as the three radiocarbon laboratories to date the Turin Shroud.

1987 *November.* A letter was sent to the archbishop of Turin by the heads of the three labs selected to carbon date the shroud asking Ballestrero to adhere to the Turin Workshop protocol and, in particular, not limit the number of labs to three.

1987 *December.* The archbishop of Turin wrote to the three labs informing them that he intended to stick by the terms of his October letter.

1988 *January.* Gove and Harbottle held a press conference on 15 January at Columbia University to protest the archbishop's abrogation of the Turin Workshop Protocol. On 22 January 1988 representatives of the three chosen carbon dating labs and the British Museum met in London, England with Gonella and agreed to the archbishop's conditions.

1988 *March.* Gove wrote to the pope outlining all that had happened and appealed to him to persuade Cardinal Ballestrero to revert to the original protocol. Gove also wrote to US Senator Moynihan to request that the US ambassador to the Vatican discuss the matter with the Vatican secretary of state.

1988 *April.* Samples were removed from the Turin Shroud on 21 April 1988 and they and control samples were given to the three laboratories.

1988 *May.* Moynihan wrote to the US ambassador to the Vatican asking him to do what he could to get the two New York State laboratories who played such seminal roles in developing small-sample carbon dating, namely Rochester and Brookhaven, involved in dating the shroud. Alas it was too late. On 6 May 1988 the first measurement of the age of the shroud's linen was made. It occurred at the University of Arizona's AMS facility. The shroud dates to the middle of the 14th century.

1988 *October.* On 13 October 1988 Cardinal Ballestrero announces the medieval date 1325 ± 33 AD for the Shroud of Turin.

1989 *February.* The results appear in the 16 February 1989 issue of *Nature*.

Index